Insect Pest Management
Ecological Concepts

Insect Pest Management
Ecological Concepts

Professor T.V. Sathe
Entomology Division
Department of Zoology
Shivaji University
Kolhapur – 416 004
&
Dr. (Mrs.) Jyoti M. Oulkar
Jyoti P.U. College
Belgaum

2010
DAYA PUBLISHING HOUSE
Delhi - 110 035

© 2010, SATHE TUKARAM VITHALRAO (b. 1953–   )
         JYOTI M. OULKAR (b. –   )
ISBN 9789351240792

Published by	:	**Daya Publishing House** **A Division of** **Astral International Pvt. Ltd.** **– ISO 9001:2008 Certified Company –** 4760-61/23, Ansari Road, Darya Ganj New Delhi-110 002 Ph. 011-43549197, 23278134 E-mail: info@astralint.com Website: www.astralint.com
Laser Typesetting	:	**Classic Computer Services** Delhi - 110 035
Printed at	:	**Chawla Offset Printers** Delhi - 110 052

PRINTED IN INDIA

Preface

There is tremendous pressure of pesticides on various ecosystem which in turn became burden to human being. There is no pure drinking water for humans even at the depth of 450 ft and pure air to breath. Pesticides lead several serious problems such as pest resistance, secondary pest outbreak, pest resurgence, killing of beneficial organisms and general kinds of pollutions. Therefore pesticidal use in pest management should be stopped or at least it should be minimized on large scale. On this background the concept of ecological pest management will play very important role in keeping environment healthy and safe for human being. Hence, this book will be stimulatory and useful guide to every environmentalists, farmers, teachers, students and scientists in the field of pest management.

T.V. Sathe

Jyoti M. Oulkar

Contents

Chapter 1
Introduction

Control of insect pests is a chronic problem in agriculture, forestry, medical and veterinary sciences since insect pests have developed resistance to almost every group of the insecticides like chlorinated hydrocarbons, organophosphorus compounds, carbamates, Dinitrophenols, organothiocynates, etc. Secondly, pesticides cause serious problems such as water, air and soil pollution; killing of beneficial organisms such as parasitoids, predators and pollinators; secondary pest out-break, pest resurgence, interruption to food webs and ecocycles, etc. This clearly indicates that there is most urgent need to find out alternative for chemical pest control strategies. Ecological control of insect pests is ecofriendly, hazard free and pollution free. Hence, it can be considered as potential and safety source of pest management and alternative for chemical control.

Temperature, humidity, photoperiod and climatic factors like biotic and abiotic play an important role in control of pest insects. Selection of appropriate control measures at appropriate time therefore, has great relevance in pest management strategies. Insects have direct or indirect relations with several ecological factors in environment. Hence, priority should be given to nature first to interact with the pest populations. Because, natural resources have tremendous power and natural power is to be utilized first in pest management. Many times, pesions can also solve the problem of pests through nature.

In India, very little attention is paid on ecological approaches of pest control. This text will fulfill the gap of demand of ecofriendly control measures in integrated pest management.

Chapter 2
Concepts of Ecological Pest Management

In ecological pest management topographical factors such as physical characteristics of the region (large water bodies, mountain ranges, streams, etc.) climatic factors (such as temperature, humidity, photoperiod, rain, wind, etc.) and biotic factors (like natural enemies, predators and parasitoids, diseases, intra and interspecific competition between the pest species) play very important role in increasing or decreasing the pest populations in the field. In past, Bhat *et al.* (1985), Gupta and Yadava (1989 a,b), Kausik and Naresh (1989), Shivankar and Kavadia (1989), Singh and Kavadia (1989), Sinha *et al.* (1989), Naritam and Sukhani (1989, 1991), Mani and Krishnamoorthy (1990), William and Ananthasubramanian (1990), Meera and Muralirangan (1990), Deshmukh *et al.* (1992), Shetgar *et al.* (1993), Veda (1993), Oulkar and Sathe (1995), Sathe (1986, 1987, 2000, 2002, 2006), Sathe and Chougale (2008),

Sathe *et al.* (1987, 2003), etc. attempted the work related to ecological insect pest management.

Role of Topographical Factors in Pest Suppression

Oceans, mountains, soil etc. affect the pest populations in natural condition. Oceans become effective barrier for many migratory pest insects. Nearly all pest species get affected by the barrier of ocean. Even smaller water bodies like lakes and large streams also become barrier to several insect pests for their life activities and spread. Mountain ranges don't allow to insect pests to cross and offer varying climatic conditions which are detrimental. Colorado potato beetle face the danger of Rocky mountains in this way. Certain flies, mosquitoes and beetles have to face the problem of characteristics of streams.

Soil feature of a region can exert a considerable impact upon the survival of the insect species. The growth of the host plants is dependant on soil and the quality of host plant affect the survival of pest insects. Soil characteristics thus, is directly or indirectly responsible for affecting pest survival. White grub beetles *Holotrichia* spp. and *Leucopholis* sp., *Anomala* sp. etc. needs fertile soil and humid climate for building the population. Hence, their population in Kolhapur region (Punchganga river area) is abundant and it is almost nil in Parbhani of Marathwada region which is relatively dry and hot humid. Some species of wire worm live comfortably in poorly drained soils but can not in clay soil.

Role of Climatic Factors in Pest Suppression

Climatic factors such as temperature, relative humidity, photoperiod, rain, wind, etc. have direct impact on control of pest populations. Only few species of insects have become adopted to variations in climate. Hence, they occur in Arctic temperate and tropical zones of the world and multiply. Thus, pest populations of a region are dependant directly or indirectly on temperature they facing and the amount of humidity they responding. In Arctic region mosquito population is increased in summer. Winter temperature control the distribution of many pests by rendering compulsion of diapousing stage or killing them due to severe cold. Harlequin bug on cabbage crop can be thus controlled. It is accepted that a very warm, moderately humid climate and fertile soil offer conditions favourable for a large number of insect pests. A limited amount of growth of a crop will produce limited pest population.

A hot and very dry climate becomes unfavourable for many pest species. Similarly, amount of sunshine affect the survival of the pests. Seldomfly and Chinch bugs are affected by the amount of sunrays. Wind velocity is powerful natural destructive source for insect pests. Many small and frailer pest insects buffeted and beaten by the wind to their death. A wheat pest belongs to the family Cecidomyidae of order Diptera is affected by this way.

Role of Biotic Factors in Pest Suppression

Biotic factors refer to parasitoids, predators, diseases, competitions etc. The pest species compets for food, mate and shelter within the species (intra specific competition) or between the species (interspecific competition).

(i) Parasitoids

Parasitoids are entomophagus insects. Their hosts are scattered in the insect world only. They always kill their host at the end of their development. Thus, parasitoids are very good biocontrol agents of insect pests. Parasitoids differ from the true parasites by having no hosts to them from non-insects and secondly, true parasite never kill its host. In total insect population, parasitoids are 15 per cent. The insect Orders Hymenoptera, Diptera, Trypsiptera, Lepidoptera, etc. are largely visualized as parasitoid producing orders. Hymenoptera ranks first by number of parasitoids. From Hymenoptera more important parasitic families refer to Ichneumonidae, Braconidae, Chalcidae, Trichogrammatidae and Eulophidae etc. The family Ichneumonidae contain 60,000 described species which are biocontrol agents of several lepidopterous, hemipterous and coleopterous pests. From family Braconidae 40,000 species have been described as pest biocontrol agents (Sathe, 2004). The family Trichogrammatidae is most famous for being egg parasitoids of lepidopteran pests in the form of _Trichogramma_ spp. and have been widely mass reared in USA, USSR, China, India, Brazil, Israel, U.K. etc. The family Chalcidae can attack larvae or pupae of insect pests and cause mortalities in the pest species. Several workers (Coppel and Martins, 1971; Sathe and Bhoje, 2000; Sathe 2004; Sathe _et al._, 2003 etc.) attempted studies related to biological control pest insects by using parasitoids are listed in bibliography of the text.

(*ii*) Predators

Insect predators are scattered in various groups of animal kingdom which refer to

1. Vertebrate predators
2. Invertebrate predators
3. Insect predators

Vertebrate Predators

Birds, small mammals, toads, snakes and fishes largely predate on insects and suppress field populations. Mammals like squirrels eat white grubs and other soil insect pests. Moles, shrews, skunks are largely dependant on insect diet. Shrews feed on pupae of lepidopteran and hymenopteran pests in forest ecosystem. Snakes, newts and salamanders subsists largely upon insect food. Toad is a very potential biocontrol agent of several pests in paddy and other agricultural ecosystems. *Bufo marinus* feed on sugarcane white grubs in Kolhapur region of Maharashtra. *Gambusia, Minnows* and *Poecilia reticulata* feed on mosquito larvae in aquatic habitat and suppress mosquito populations on large scale. Bats feed on several pest moths of economic importance.

Birds have immense value in insect pest management. A birdless country will be the most desirable place for insects. Robins and catbirds mostly feed on insects in summer season while, wood peckers and creepers enjoy insect diet throughout the year. Similarly, English sparrow and black bird largely subsists on insects. Indian mynah, *Achriotheres tristis* is manageable in control of locusts and several other insects. Waigtail bird feed on flying mosquitoes at evening.

Invertebrate Predators

Insect pests are predated by several invertebrate predators rather than the insects. Spiders, mites and nematodes suppress several populations of pests. Mermicid worms (Nematoda) are exclusively parasitic on insects such as grasshoppers, locusts, fruit flies, elm bark beetles etc. Chlorohydra predates on mosquito larvae and reduce the populations of mosquitoes. Planarians control mosquito larvae in aquatic habitat.

Insect Predators

Insect predators constitute major factor in the environmental resistance. Order Coleoptera and Neuroptera constitute large number of insect predators. Hemiptera, Hymenoptera, Lepidoptera also provide pest predatory insects. The family Coccinellidae of Order Coleoptera is quite famous for predatory insects. Coccinellid beetles or ladybird beetles comes under this family. Upto date 4000 species of lady bird beetles have been described from the world (Sathe and Bhosale, 2001). They feed on aphids, jassids, whiteflies, scales, mealy bugs, psyllids etc. and suppress their populations in the field. Tiger beetles, Cicindelid beetles also play an important role as natural enemies of several insect pests. Neuropteran Lacewings are also very potential natural enemies of insect pests. *Crysoperla carnea* is very potential natural enemy of sugarcane wooly aphid in Maharashtra. In natural population dragonflies and praying mantids have greater role to play. They suppress populations of jassids, aphids, grass-hoppers, small moths, mosquitoes, midges, etc. More attention is to be paid on these important groups of natural enemies for ecological pest management.

Ecological Pest Control in some Pests

The gram pod borer *Helicoverpa armigera* (Hubner) (Lepidoptera : Noctuidae) is serious pest of pulses which is responsible for heavy crop losses. Yadava and Lal (1988) have made the survey of environmental factors that regulate the cyclic occurrence in 35 districts of Uttar Pradesh. They reported two larval peaks in the pest; first between 47–50 standard week and between 11–15 standard week. The correlation coefficients were found to be positively significant in case of maximum and minimum temperatures but were negatively non-significant with relative humidity and per cent parasitization. Kaushik and Naresh (1989) worked on sampling techniques for different insect pests. There is paucity of information on suitable sampling procedures, of great importance in pest ecology.

Gupta and Yadava (1989) studied component of natural enemies on the pest aphid. *Myzus persicae* (Sulzer) (Hemiptera : Aphididae). This is serious pest of Cumin from flowering to maturity. Five species of Coccinellid predators *viz.*, *Coccinella septumpunctata, Coccinella* sp., *Brumoides suturalis, Menochilus sexmaculatus* and *Adonia variegata* and one predatory bird Indian mynah, *Achriotheres tristis* were recorded predating upon the aphid, *M. persicae* in the Cumin field. The maximum aphid population and its rate of change was recorded when predators were excluded from the crop by screening and minimum was under natural condition which signified a positive role of predators in regulating the aphid population. Encouragement to predators is very good approach of ecological pest management of aphids.

To minimise the dependence on use of insecticides for insect control ecofriendly control like alteration in sowing date may be useful. Insecticides cause environmental pollution which makes life more miserable and increases the danger of poisoning of mankind as well as of the useful animals. There is least incidence of aphid on early sown crop *i.e.* the crop sown upto 16th November. Late sown crops are progressively more severely infested. Higher yields are obtained in the crop sown on 1st and 16th November and the crop sown after these dates have drastic reduction in yield.

The residual toxicity and persistence of insecticides are dependent on the multitude of environmental factors. The deterioration in the toxicity of carbaryl, endosulfan and malathion was quicker when treated surfaces were exposed to sunlight than kept in the shade. The dissipation of insecticides was faster when the treated surfaces were subjected to 8 hours sunlight and 16 hours shade per day than when subjected to continuous 24 hours shade condition (Singh and Kavadia, 1989). The greater loss of insecticides is possible under sunlight having intense ultraviolet radiation and high temperature. Greater loss of DDT and its derivatives due to decomposition volatility was also due to exposure to sunlight. Fleek (1948) also reported that the residual action of non-volatile organic insecticides was governed by their resistance to chemical change under field conditions. Oxidation has been found to be a common cause of failure of pyrethrum, rotenone, phenothiazine and DDT. These actions are catalized by light.

Lipaphis erysimi Kalt (Hemiptera : Aphididae) is serious pest of mustard which reduce the yield of crops as about 80 per cent. The ecological approach to the pest management suggest to use pesticide only when and where necessary. This is possible only after making survey of ecological factors interacting with pest populations. According to Singh *et al.* (1969) the aphid was found to appear and establish on *Brassica* spp. in the third week of December. It build up its population in January–February reaching the peak on 8th and 18th February in 1980 and 1981, respectively. The environmental parameters played an important role in the aphid infestation. The ambient maximum (21.68–23.52°C) and minimum (7.18–9.40°C) temperatures in January–February appeared to be most conducive for the aphid multiplication. High humidity had little impact on aphid population fluctuation. Minimum humidity ranging from 55.7–69.4 per cent in January–February favoured the population build up. While, the activity of the aphid ceased at 50.90 per cent and below. Frequent rains during the population rise phase (January–February) adversely affected in the aphid population. Keeping if low through the period of its abundance in 1980–81, Roy (1976) also reported similar observations in the pest which would be useful for minimizing the use of pesticides when biocontrol agents interacts with pest populations.

It is now accepted fact that application of ecological principles to crop protection is a sound strategy of pest management. Insecticide residues in plant sap during susceptible stage of seedling must be maintained at a level of which should be toxic enough to kill larvae inside the stem. Naitam and Sukhani (1989) studied the effect of soil, weather and other crop growth factors on pesticidal efficacy against sorghum shoot fly *Atherigona soccata* (Rond.) and reported that sowings of carbofuran treated and un-treated seed in different seasons have indicated carbofuran improved seed germination under sufficient rainfall conditions (Kharif season) but. germination was very little during summer and rabbi seasons. The rainfall has also the relationship with oviposition. More rainfall resulted into lesser oviposition in sorghum shootfly. Likely, more soil moisture resulted into lesser shootfly damage and increased the yield.

In case of light sandy soil, the insecticides proved ineffective against the shootfly when there were heavy rains. In such conditions the insecticides are washed off from the site of application and leached down into the soil thereby reducing the uptake by the plants. The environmental factors like rainfall and soil moisture have important bearing both on rate of plant growth and effectiveness of systemic insecticides when applied in soil. The insecticidal efficacy various from place to place and as such carbofuran, a most potential insecticide did not prove equally effective against sorghum shootfly at several places. Therefore, Naitam and Sukhani (1991) studied effect of soil, weather and other crop growth factors on the insecticidal efficacy against sorghum shootfly *A. socata* and reported that the efficacy of soil applied insecticides was highly affected by soil properties and environmental conditions during different seasons. However, carbofuran improved seed germination and seedling growth but this boosting effect was evident only under specific conditions. The insecticidal efficacy in the black cotton soil of Parbhani (M.S.) was comparatively poorer than that in the sandy loam soils of Delhi region. The characteristics of soil and weather affect the efficacy of pesticides resulting in control of pest species and increase in the yield of crops.

The mango hopper *Amritodus atkinsoni* (Leth.) (= *Ediocerus atkinsoni* L.) is one of the most serious pest of mango trees in Gujarat and Maharashtra. The hopper cause 25–60 per cent damage to crop by feeding on inflorences and developing fruits by sucking the cell sap. Its' sex ratio and parasitization behaviour have been studied with respect to temperature and humidity in south Gujrat by Patel *et al.* (1989). They reported that the males and females found to have preferences for region of mango trees. The females preferred upper part of the mango tree from February to April and September to November. But, both sexes migrated during April, May, June and July to the lower part of the tree. According to Bursell (1964) fecundity has close relation with humidity. From May to September the humidity ranged from 76–90 per cent which was very high than other months of the year, was responsible for inactive siting behaviour and non-productive behaviour. This ecological behaviour is manageable for adopting special control strategies.

The sunflower (*Helianthus annus* L.) in Uttar Pradesh is widely cultivated (23,333 ha). The faunastic survey of insects on sunflower has been already carried out in various states. Goal and Kumar (1989) studied the seasonal build of insect pests on a monsoon crop of sunflower in an agro-ecosystem of Muzaffarnagar. The analysis of dominance of certain pests with respect to weather factor will add great relevance in adopting ecological pest control strategies against these pest species. The phenological studies revealed a vital impact of the minimum temperature and relative humidity for insect abundance in a monsoon crop of sunflower (EC 68414). Species of major numerical dominance associated with different parts of the crop were *Attractomorpha crenulata* (F.), *Leptocentrus taurus* F., *Myllocerus lateralis* Chev., *Sarcopleaga* sp., *Carpophilus* sp., *Brumus suturalis* (F.). *Camponotus compressus* F., and several lepidopteran caterpillars. The average agroclimatic factors of the cropping period were 34.6±0.22°C max., 26.5±0.42°C min. temperatures and 66.2±2.83°C relative humidity. In all 458.5 mm rainfall was recorded during June to August, 1979. Of the Orthoptera associated with the monsoon crop of sunflower, 79.6 per cent of the total individuals were *Crenulata*. Odonata and mantoidea were groups of predaceous insects in addition to Lady bird beetles.

Cowpea *Vigna unguculata* (L.) Walp. is one of the important pulse crops of the tropics and is infested by number of insect pests which constitute a serious constraint in realising the potential yield. The weather conditions prevailing in a region play an important role in the occurrence and build up of insect pests of cow pea and their biotic agents from Delhi region. *Aphis craccivora* Koch population and minimum temperature showed a significant negative correlation. In *Plusia orichalcea* F. maximum daily temperature and sunshine had a significant negative association with the build up of population. Faleiro *et al.* (1990) showed that the relative humidity and rainfall were significantly and positively associated with the build of the pest. For *Aphacospora* sunshine had a significant negative influence on the build up of population. The maximum daily temperature showed significant negative association with the spider population. The maximum daily temperature had a significant positive association bearing on the build up of yellow wasp (*Polistes herbreus* F.). Maximum daily temperature and sunshine were significant and positively influenced the build up of *Brumus* sp. population. However, the build up of the lady bird beetle was not significantly influenced by any of the weather parameters. According to Singh *et al.* (1990). fairly high mean ambient temperature around (32–34°C) seems to be most conducive in the population build up of various insect pests of groundnut. The studies made by Mani and Krishnamoorthy (1990) on natural suppression of mealy bugs in guava orchards revealed the presence of one parasitoid and four predators on the citrus mealy bug *Plamococcus cilti* (Risso) and a parasitoid and one predator on the striped mealy bug *Ferrisia virgata* (Ckll.). The green lace wing, *Chrysopa lacciperda* (Kimmis) the lycaenid *Spalgis epius* Westwood and the coccinellid *Cryptolaemus montrouzieri* Mulsent reduced the population of *P. citri*.

Linseed *Linum usitatissium* L. is an important oil seed crop of India. India cultivates nearly 1/4 per cent of its total world acreage and ranks first in area and fourth in production in the world. However, expected yield of the crop is not achieved so far because of damage caused by various insect pests. Therefore, Deshmukh *et al.* (1992) studied pest complex and their succession on linseed for hopping the control of insect pests by ecological means. They reported that the population of different pests occurred in an overlapping manner and that the crop was continuously infested with one or the other pests. Patterns of pest incidence on this crop however, revealed that *Agrotis ipsilon* (Hfn.) was a potential threat to the crop during the vegetative phase, while *Dasineura lini* Barn., *Helicoverpa armigera* (Hubn.) and mites (*Petrobia latens* Mull, *Byrabia* sp.) proved menancing to the crop during reproductive phase. *Phytomyza horticula* Gour was the only pest which infested the crop throughout its growth in vegetative and reproductive phases. They reported natural enemies of the pests, Lady bird beetle (*Coccinella septumpunctata* L.), Lace wing (*Chrysopa scelestes* Bank), yellow wasp (*Polistes breaus* F.) etc. Encouragement to natural enemies and minimization of use of pesticides is very good approach of ecological pest management in above case.

The climatic factors play substantial role in the biology of any pest of which the temperature is most crucial abiotic factor influencing the life economy of any organism. It is rather difficult to find a direct cause and effect relationship between any single climatic factor and pest activity because, the impact of weather elements on pest is usually confounded. However, temperature, rainfall, relative humidity, sunshine and wind speed are the chief weather parameters that largely direct the activity of a given species of insect. Jayanthi *et al.* (1993) studied population build up of insect pests on MH-4 variety of groundnut influenced by abiotic factors. They reported that the ambient temperature around 35.5°C, fairly high relative humidity (71.6 per cent), low wind speed (3.9 / km/hr) and a long sunshine duration (8.3 hr/day) influenced the population build-up of insect pests which attained their peaks in an overlapping manner. The key pest *Empoasca kerri* Pruthi showed a positive correlation with maximum daily temperature, sunshine, rainfall, evening relative humidity and negative relationship with minimum daily temperature, morning relative humidity and wind speed. The population of *Caliothrips indicus* Bagnall had a positive correlation with temperature, sunshine, rainfall, morning relative humidity and a negative correlation with evening relative humidity and wind speed. The population of *Bemisia tabaci* (Genn.) showed a positive association with temperature, rainfall, wind speed, evening relative humidity and a negative correlation with morning relative humidity and sunshine. The abiotic factors as independent variables and the population of jassids thrips, whiteflies as dependent variables showed significant effects on above pests of groundnut.

Pigeon pea *Cajanus cajan* (L.) Millsp is an important pulse crop of India. It is grown as sole crop as well as mixed and inter crop in India. Gram pod borer *Helicoverpa armigera* (Hubn.), plume caterpillar *Exelastis atomosa* Walsingham, pod fly *Melanagromyza obtusa* Mall. and tur pod bug *Clavigralla gibbosa* Spin. are major insect pests which cause severe losses to the pigeon pea crop. Veda (1993) studied the effect of weather factors on the incidence of pod bug, *C. gibbosa* in pigeon pea in Madhya Pradesh. He reported that weather factors play a key role in the development of pests. The pest started its activity from first fortnight of September and continued till the harvest of crop. The population ranged between 2.3 to 45.3 nymphs and adults per 2.5 m² cropped area. Both temperature and rainfall played a key role in the multiplication of pest. Maximum mean fortnightly temperature of 32.33°C and minimum of 27.26°C coupled with a rainfall of 98.50 mm during the October provided the congenial conditions for the development of this pest. The damage to pods and grains of pigeon pea showed a loss to the tune of 25.2 and 20.38 per cent respectively.

Parasitoids have very important role in ecological pest management. Parasitoids cause mortalities in egg stage, larval stage, pupal stage and even adult stage of the pest species. In general, eggs are deposited into the host body. After hatching the eggs, the newly emerged larvae feed on internal tissues of the host and develop into certain instars (usually 3–5). Finally, the last instar break the body wall of pest insect and come out by killing the pest and prepare cocoon outside the body of host and pupate inside the cocoon. Lastly, the adult come out of the cocoon and the parasitoid again search for pest for egg laying and for its own development. Of the total biocontrol programmes designed in the world, in 60 per cent programmes, parasitoids have been used successfully.

Groundnut leafminer *Aproaerema modicella* Deventer is one of the major insect pests of groundnut in India. Natural control potential of parasitoids of this pest has been studied by Shetgar *et al.* (1993) in Maharashtra. In all, 7 parasitoids *viz. Goniozus stomopterycis* Ram. and Subba Rao, *Chilonus* sp., *Dolkshogaster* sp., *Stenomesius japonicus* (Ashmead). *Sympiesis dolichogaster* (Ashmead), *Eurytoma* sp., and *Anagrus* sp. were found to be parasitizing groundnut leafminer. The larval mortalities were observed to be 16.36, 25.00 and 26.23 per cent because of the 7 different parasitoids in 3 generations, respectively. The mortality factors operating during first generation were not effective in reducing the pest population in second generation. However, the mortality factors operated during second generation were effective in bringing down the pest population in third generation.

Insect predators of pests also have tremendous importance in ecological pest suppression. From family Cicindellidae of order Coleoptera 20,000 species have been reported as insect predators. The family coccinellidae contain 4000 described lady bird beetles. *Rodalia cardinalis* is the first lady bird beetle used in biological control of the pest insect, *Icerya purchasi* in western countries. Lady bird beetles cause mortalities in several insect pests of sub order Homoptera (soft bodied individuals) including aphids, jassids, mealy bugs, white flies etc.

Mealy bug *Maconellicoccus hirsutus* (Hemiptera : Pseudococcidae) is very serious pest of grapes, mulberry, mango and several horticultural crops. Australian Lady bird beetle *Cryptolaemus montrouzieri* Mulsant is very potential biocontrol agent of mealy bug. The efforts have been made to control the mealy bug with pesticides but not yielded desired results. Biological control, the next important method to chemical means was attempted by releasing Australian lady bird beetle in the grape gardens of Hyderabad region but, this predator couldn't establish in the region due to indiscriminate use of pesticides in grape ecosystems. Therefore, Subbaratnam *et al.* (1992) tested toxicity of some pesticides against this beetle. They tested carbaryl 0.15 per cent, phosalone 0.07 per cent, chloropyriphos 0.05 per cent and sulphur 0.1 per cent sprays. Their results indicated that chloropyriphos 0.05 per cent was least toxic to the beetle C. *montruizieri.* Carbaryl 0.05 per cent was highly toxic to the predator and toxicity persisted 27 days in the field. Similarly, phosalone 0.07 per cent was also highly toxic to the beetle and toxicity persisted for 23 days in the environment. As like chloropyriphos, sulphur was also less toxic to lady bird beetles and safe period detected for beetles was 5 days. Thus, avoiding the use of pesticides where biocontrol agents are working and using safer pesticides to biocontrol agents is ecofriendly strategy in integrated pest management.

Photoperiod is an important abiotic factor affecting growth, development and metabolism. It also influences the survival capacity of nymph and adult longevity. In a grasshopper *Oxya nitidula* (Walker) egg maturation, fecundity, egg mortality and adult life span are very much influenced by changes in photoperiod. The high rate of female fecundity, observed under LL regime, indicates the influence of light in stimulating greater egg production. In *Anacridium aegyptium* rapid synthesis and transport of neurosecretary materials of brain in reproductively active females and slow accumulation of cysteine material in the female with reproductive arrest have been demonstrated (Geldiay, 1970). The duration of embryonic period does not show much variation in certain insects. However, the post embryonic development of male is greatly delayed in the LL regime and still more under DD regimes. The longer photophase has a positive effect on the nymphal survival of *O. nitidula*. The stimulus provided by the change in the duration of light is probably clock. Biological dock of the insect will tell us the emergence rhythum which can be in turn manageable in appropriate ecofriendly control measures.

Termites are quite destructive pests of agriculture and forestry. There are about 2000 species of termites in the world. Termites are social insects with polymorphism, caste system and division of labour and great concern for ecological studies to be managed in pest control strategies. On the ecological basis termites are of three kinds:

1. Moist wood termites
2. Subterranean termites and
3. Dry wood termites

Of the above three kinds subterranean termites are most important because they cause severe damage to our agriculture and forest properties. The termite can digest cellulose hence, they feed on woody content in various ecosystems. Roonwal (1988) studied the field ecology in Indian Sand termite *Psamotermes rajasthanicus* Des (Isoptera). Accounts of its ecology and pest status, as attacking wood work, poles and forest trees etc. have been reported by Roonwal (1968–78). This species occurs in arid areas characterized by open scrub and loose, sandy soil, and is common under stones, semi-dry cowdung etc. This species also reported from below the bark of dried up trees and in dead wood. It can occur with other termite species like *Microtermes mycophagus* where it some time attacks wood work in rural houses and wooden poles. In western

Rajasthan it attacks some forest trees as saplings and 6–10 year old plants (*Tecomella* sp.). It also attacks *Capparis decidua*; sheesham *Dalbergia sissoo, Acacia* spp., *Tamarix* spp., *Zizyphus jujuba*, etc. Nesting occurs undergound quite deep in soil. It prefers low rainfall. It is found in very arid western zone of Rajasthan. Its natural enemies, inter and intra specific competition and weather factors are awaiting and would worthwhile in adopting its ecological pest control. Termites of Western Ghats have been studied by Sathe and Chougule (2008) with respect to diversity, habitat, seasonal abundance and host plants damaged. Rains always suppress the termite populations.

Competitions amongst the pest species is also of great concern to ecological insect pest control strategies. Competition in pest species is for food, mate and shelter. The competition may be inter-specific or intra-specific for above basic needs. The insects can be controlled by depriving them from their food, mate and shelter. The coding moth is serious pest of apple tree, is controlled by scrapping the wood bark of apple trees where they hybernates in last instars. Cockroach needs warmth and food. These can be avoided by keeping the kitchen cool and providing no food. Mantid males fight between themselves for mate and cause mortalities and/or suppress their own population and in turn they interacts with low number with pest populations. Same thing is happened with dragonfly predators.

Cannibalism is phenomenon in which the individual species feed on their own fellows and suppress their population in the nature. Cannibalism is good example of intra-specific competition for food. Cannibalism is reported in gram pod borer *Helicoverpa armigera* (Sathe *et al.*, 1988),

Flour beetle *Tribolium confusum,* codling moth *Cydia pomonella* etc. Oulkar and Sathe (1995) studied the Cannibalism in a hairy caterpillar *Amsacta lactinea* (Cramer) wherein they found that cannibalism was highest in older instars than young instars due to want of food.

Diseases of insect pests have tremendous importance in ecological pest management. Fungi, bacteria, viruses, rikettissae, protozoans and Nematodes are important pathogens of pest species. Fungi contains 750 entomogenous species. Bacteria have more than 90 entomogenus species and viruses about 450. Protozoan contains 210 insect parasitic species while, Nematodes have 1000 species parasitic on insect pests. Utilization of pathogenecity in insect pest control has great relevance from the view point of ecological control of insect pests.

Chapter 3

Rearing of Pests and Parasitoids

For reducing the pest populations, control measures are needed. In certain pest control measures, continuous breeding stocks of pests are necessary. The biocontrol methods are more efficient methods continuously being adopted and are on going processes. Some times very minor change in techniques can have drastic results, both positive and negative. Hence, in the present study, following material and methods were adopted for the experiments.

1. B.O.D. Incubator (Figure 1)
2. Air Cooler (Figure 2)
3. Glass cages (Figures 3 and 4)
4. Glass Troughs (Figure 5)
5. Plastic containers (Figure 6)
6. Petridishes (Figure 7)
7. Test tubes
8. Specimen tubes

Plate I

Figure 1: B.O.D. Incubator; Figure 2: Air Cooler; Figure 3: Glass Cage Type–I; Figure 4: Glass cage Type–II; Figure 5: Glass Trough; Figure 6: Rack with Plastic Containers; Figure 7:: Petridishes.

Plate II

Figure 8: Scorpion Caterpillar; Figure 9: Scorpion Caterpillar; Figure 10: Lymandsid Caterpillar; Figure 11: Unidentified on Til; Figure 12: Satyrid Caterpillar–*Melantis* sp.; Figure 13: *Amsacta* sp.; Figure 14: *Amsacta* sp.; Figure 15: Unidentified on Soyabean; Figure 16: *Spilosoma oblique;* Figure 17: *Amsacta* sp.; Figure 18: *Thiocidas postica.*

Plate III

Figure 19: *Eupterote fabia* Cramer; Figure 20: *Spilosoma obliqua* Walker; Figure 21: *Sangatissa* sp.; Figure 22: *Nisaga* sp.; Figure 23: *Amsacta* sp.; Figure 24: *Syntomis cyssea* Cramer; Figure 25: *Syntomis* sp.; Figure 26: *Syntomis cyssea* Cramer; Figure 27: *Nisaga simplex* Walker; Figure 28: *Spilosoma obliqua* Walker; Figure 29: *Amsacta* sp.

Plate IV

Figure 30: Eggs of *S. oblique;* Figure 31: First Instar Larvae of *S. obliqua;* Figure 32: Second Instar Larva of *S. obliqua;* Figure 33: Third Instar Larva of *S. obliqua;* Figure 34: Pupa of *S. obliqua;* Figure 35: Adult of *S. obliqua;* Figure 36: Eggs of *A. lactinea;* Figure 37: Larvae of *A. lactinea;* Figure 38: Pupa of *A. lactinea;* Figure 39: Adult of *A. lactinea;* Figure 40: Eggs of *T. postica;* Figure 41: First Instar Larvae of *T. postica;* Figure 42: Second Instar Larvae of *T. postica;* Figure 43: Third Instar Larvae of *T. postica;* Figure 44: Fourth Instar Larva of *T. postica;* Figure 45: Pupa of *T. postica;* Figure 46: Adult of *T. postica;* Figure 47: Infected *S. obliqua* Larva; Figure 48: A Sunflower Field and Author.

Plate I

Plate II

Plate III

Plate IV

Methods

Rearing of the Species

Rearing of *Spilosoma obliqua* (Plate IV Figures 30–35)

Rearing of *S. obliqua* was carried out under laboratory conditions (21±1°C, 55–60 per cent R.H., 12 hr. photoperiod) by collecting the larvae (Figure 31–33) from the fields of soybean and sunflower. For pupation (Figure 34) larvae were reared in glass troughs. For adult (Figure 35) emergence, newly formed pupae were kept in separate containers. After the emergence of moths, a pair (♂:♀) was kept in glass jar for mating. After mating, female started egg (Figure 30) laying on tissue paper which was wrapped internally in jar and on the wet muslin cloth which was soaked in water. Female lays about 352 to 457 eggs with the help of ovipositor, the egg mass was separated by cutting the paper and placed in petridishes. Newly emerged larvae (Figure 31) were placed in container. Later instars were transferred separately into small containers (Figure 32) to avoid cannibalism. During developmental period 50 per cent honey and sunflower leaves were used as food for moths and larvae respectively.

Rearing of *A. lactinea* (Plate IV Figures 36–39)

For the rearing of this species, the laboratory conditions were 25±1°C, 55–60 per cent RH and 12 hr photoperiod. The larvae (Figure 37) of *A. lactinea* were collected from the fields and reared individually in plastic containers for pupation (Figure 38) and adult (Figure 39) emergence. Newly emerged pair (♂ and ♀) of moths was kept in glass jar for mating. After mating female laid about 199 to 288 eggs on tissue paper and wet muslin cloth of jar. Eggs

(Figure 36) were collected by cutting the paper/cloth piece and placed in petridishes. In each container, 50 newly emerged larvae (Figure 37) were taken and kept for further development. After IInd instars the larvae were kept separately in containers. The larvae were fed with groundnut leaves and adult moths with 50 per cent honey.

Rearing of *T. postica* (Plate IV Figures 40–46)

Rearing of *T. postica* was carried out under laboratory conditions (25±1°C, 55-60 per cent R.H., 12 hr photoperiod) by collecting the larvae of *T. postica* from the fields. Larvae were reared in glass troughs for pupation (Figure 45). Newly formed pupae were kept in separate containers for adult (Figure 46) emergence. After the emergence of moths a pair (♂ and ♀) was kept in glass jar for mating. After mating female laid about 287 to 319 eggs on tissue paper and on muslin doth. The eggs (Figure 40) were collected and placed in petridishes. Newly emerged larvae (Figure 41) were kept in plastic containers and after IInd instar they were transferred separately into small containers for further development. 50 per cent honey and ber leaves were used as food for moths and larvae respectively during the experiments.

Rearing of Parasitoid Species

Newly emerged parasitoids (♂ and ♀) were caged into test tube (size 15 × 2 cm) for their mating. The mating was followed immediately in test tube. For parasitization, early second instar larvae of the hosts were exposed in glass cages (Figures 3, 4) to 5 mated females of parasitoids. The host larvae were introduced in cage through sleeve with the help of fine hair brush. More parasitized host larvae were obtained with a short time since one or other parasitoid quickly oviposited into host larvae offered. Parasitization was also made in test tubes (size 19 × 2.5 cm). Host larvae were removed after parasitization and kept separately in containers for further development. The parasitoids and the hosts were fed with 50 per cent honey solution and host food plant leaves respectively during the experiments.

Chapter 4

Ecological Aspects of Lepidopterous Pests

Survey and Surveillance

Introduction

Determination of reasonably precise estimate of the number present in insect population is one of the most difficult tasks. For sampling insect populations basic knowledge of biological peculiarities of insect should be known. The specialized habits such as occurrence, survival, abundance, natural enemies, etc. are the fundamental and important aspects by which sampling of insect populations can be strengthened. Sampling of insects with specialized habits like parasitoids is a process of sampling the parasitoid hosts. For the population dynamics and for constructing the life tables, the survey of species and successive estimates of the number of insects per unit of land area are necessary. In general, the survey and surveillance of the insect has immense value for monitoring notorious pest species and, thus, it is most important strategy in the pest management. The validity of such finding is a area specific and hence, it has its own importance. Such studies will add knowledge of pest species behaviour in relation to occurrence, abundance and its natural enemies. Keeping in view the above facts, the present topic has been enlighted on some hairy caterpillars of economic importance of agricultural crops, namely *Spilosoma obliqua* (Walker), *Amsacta lactinea* (Cramer) and *Thiacidas postica* (Walker).

Materials and Methods

During the period of 1988 to 1990 extensive survey was made on the hairy caterpillars namely *S. obliqua*, *A. lactinea* and *T. postica* by collecting the caterpillars from the fields of soyabean (*Glycine max* L.), groundnut (*Arachis hypogea* L.) and ber (*Ziziphus mauritiana* L.) (Figure 49). Pupae and adults were also collected as far as possible. Weekly collections were made for 1 man 1 hr search at each field. Within the same field the sample area was changed in successive collections. The observations on their abundance were continued in the same field till the harvest of crops. The larvae and pupae collected from the fields were reared in the laboratory (25±1°C, 55–60 per cent RH, 12 hr. photoperiod) on their respective host food plants till the adult emergence. During the developmental time, observations were made on the possibilities of parasitoids emergence from the pest species, and also on the other mortality factors. The percent-parasitism was calculated by number of host larvae collected and number of parasitoids emerged from hosts. The mortality caused by other factors calculated by counting the dead larvae and

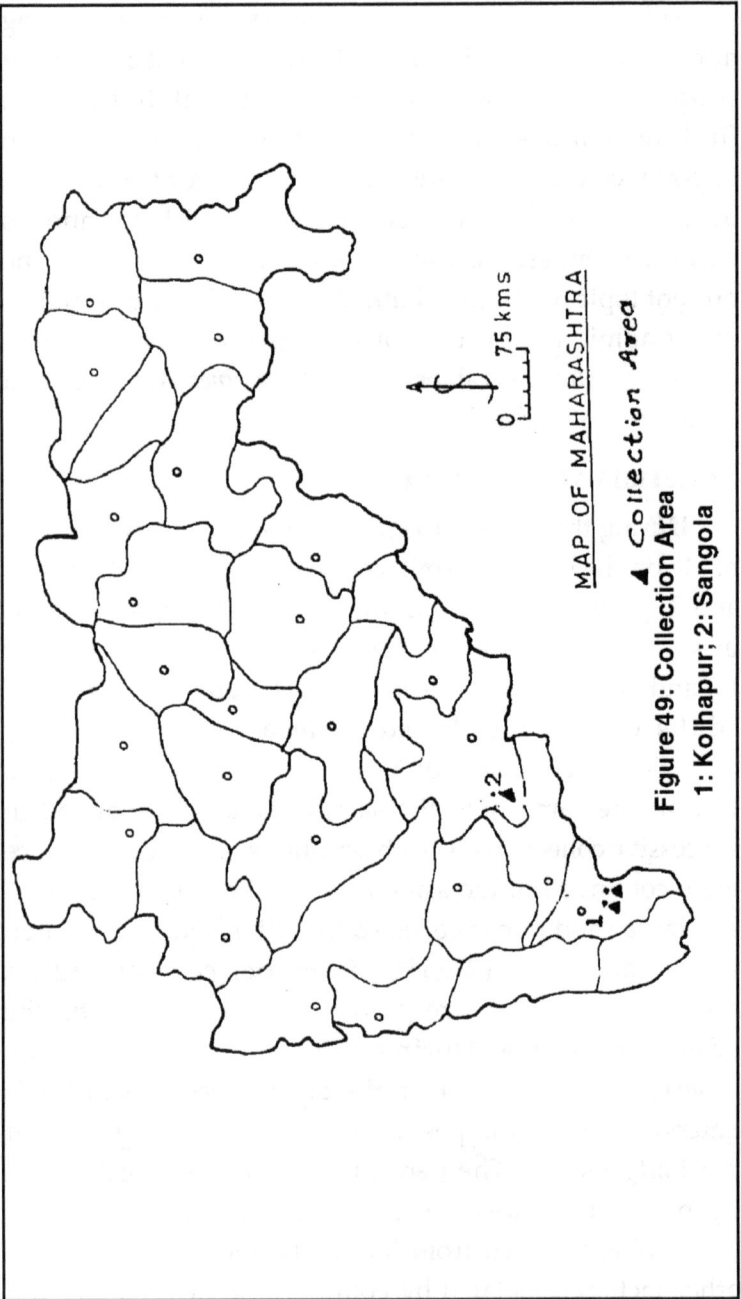

MAP OF MAHARASHTRA

▲ Collection Area

Figure 49: Collection Area

1: Kolhapur; 2: Sangola

pupae of the host during the course of studies. The meterological data mainly temperature, relative humidity and rainfall observed during the courses of investigation (from 1988 to 1990) is represented in Figures 50–52.

Results

S. oblique

The results on the survey studies of *S. obliqua* and its parasitoids have been recorded in Tables 1 to 3 and Figures 53–55. The survey of *S. obliqua* and its parasitoids were made on the soyabean from June to December. The appearance of the pest was started from the 4th week of June and reached in its peak in August and October. Parasitoid appearance was started from the 2nd week of July and later increased in the successive months (Figure 53). The intensity of pest population were considerably high during the survey studies. However, the population was greatly reduced approaching the time of harvestation of the crop.

A. lactinea

The survey of *A. lactinea* and its parasitoids was made on groundnut from June to November. The population of the pest was found increased in August to September and later it decreased upto certain extent. The pest appearance was started on groundnut from the 1st week of July and parasitoids appeared in 4th week of the July. The population of the pest was optimum during September 1st week and later found decreased (Figures 53–55). The parasitoids reared on *T. postica*, their biological peculiarities have been discussed in a separate chapter, "parasitoids" in the text. The results on survey of this pest are tabulated in Tables 4 to 6 and Figures 53–55.

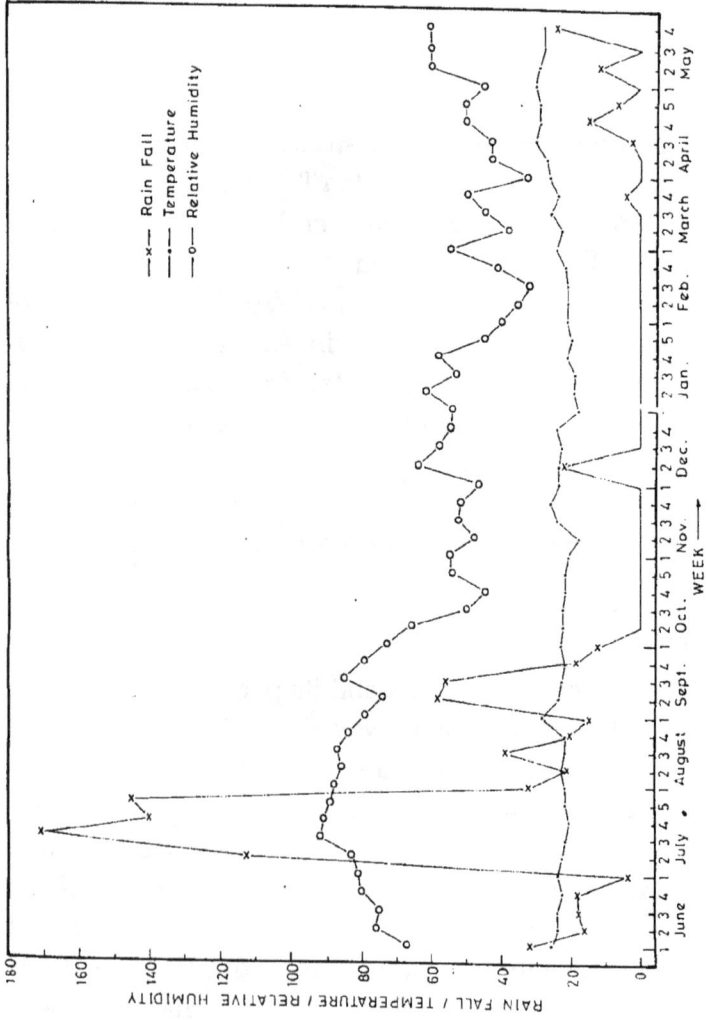

Figure 50: Meterological Data Collected at Kolhapur during Weekly Survey Study for Year 1988–89

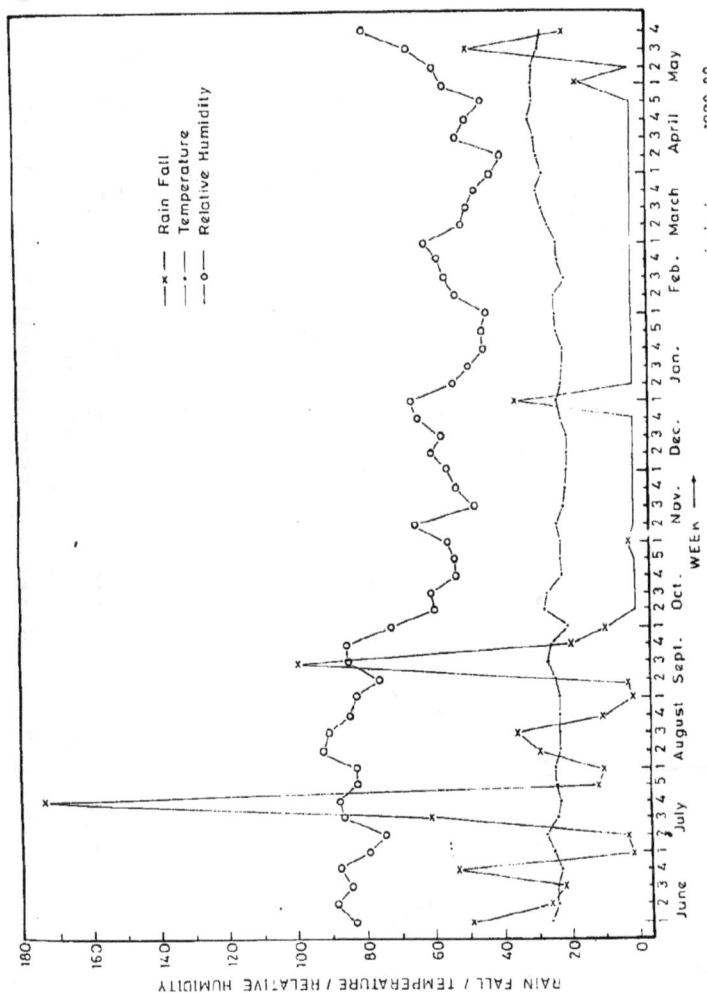

Figure 51: Meterological Data Collected at Kolhapur during Weekly Survey Study for Year 1989–90

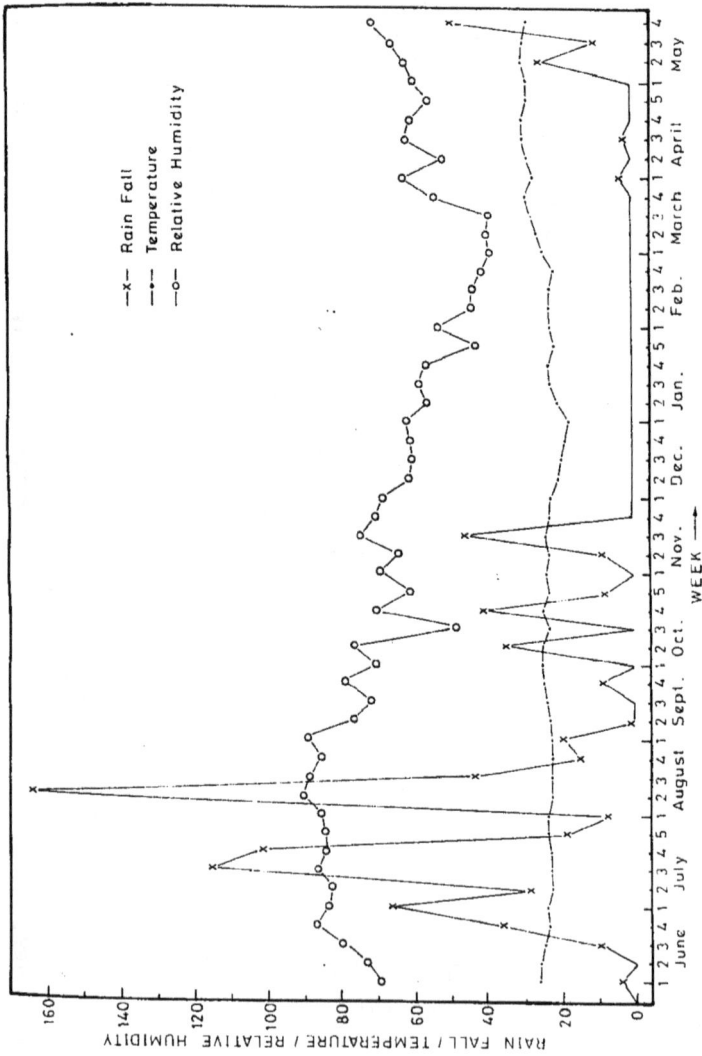

Figure 52: Meterelogical Data Collected at Kolhapur during Weekly Survey Study for Year 1990–91

Table 1: Survey of *S. obliqua* and its Parasitoids on Soyabean (1988)

Months	Date	No. of Larvae Collected	Moths	Parasitoids	Per cent Parasitism	Mortality Due to Other Factors
June	5	–	–	–	–	–
	12	–	–	–	–	–
	19	–	–	–	–	–
	26	121	73	31	–	–
July	3	141	77	36	–	19.85
	10	100	73	16	16.00	11.00
	17	–	–	–	–	–
	24	102	74	12	11.76	15.68
	31	–	–	–	–	–
August	7	204	84	98	48.03	10.78
	14	292	112	140	47.94	13.69
	21	281	112	131	46.61	13.52
	28	400	190	151	37.75	14.75

Contd...

Table 1–Contd...

Months	Date	No. of Larvae Collected	Moths	Parasitoids	Per cent Parasitism	Mortality Due to Other Factors
September	4	402	163	182	45.27	14.17
	11	388	180	156	40.20	13.40
	18	382	178	157	41.09	12.30
	25	547	235	251	45.88	11.15
October	2	502	142	289	57.56	14.14
	9	506	130	308	60.86	13.43
	16	547	152	317	57.95	14.25
	23	546	157	318	58.24	13.00
	30	551	162	317	57.73	13.06
November	6	552	266	205	37.13	14.67
	13	542	232	221	40.77	16.42
	20	536	222	232	43.28	15.29
	27	551	254	208	37.74	16.15

Table 2: Survey of *S. obliqua* and its Parasitoids on Soyabean (1989)

Months	Date	No. of Larvae Collected	Moths	Parasitoids	Per cent Parasitism	Mortality Due to Other Factors
June	4	–	–	–	–	–
	11	–	–	–	–	–
	18	–	–	–	–	–
	25	109	57	20	–	30.83
July	2	120	59	24	16.21	32.43
	9	148	76	24	20.43	24.08
	16	137	76	28	–	14.89
	23	47	40	–	21.34	23.03
	30	178	99	38	18.18	30.00
August	6	220	114	40	52.44	21.03
	13	347	92	182	57.26	19.83
	20	358	82	205	46.99	23.16
	27	449	134	211		

Contd...

Table 2–Contd...

Months	Date	No. of Larvae Collected	Moths	Parasitoids	Per cent Parasitism	Mortality Due to Other Factors
September	3	448	130	209	46.65	24.33
	10	500	383	–	–	30.54
	17	–	–	–	–	–
	24	106	59	16	15.09	29.24
October	1	320	74	167	52.18	24.68
	8	318	64	178	55.97	23.89
	15	220	128	41	18.63	23.18
	22	198	121	39	19.69	19.19
	29	178	121	36	20.22	11.79
November	5	175	117	36	20.57	12.57
	12	110	78	21	19.09	10.00
	19	95	70	11	11.57	14.73
	26	92	75	10	10.86	7.60

Table 3: Survey of *S. obliqua* and its Parasitoids on Soyabean (1990)

Months	Date	No. of Larvae Collected	Moths	Parasitoids	Per cent Parasitism	Mortality Due to Other Factors
June	3	–	–	–	–	–
	10	–	–	–	–	–
	17	–	–	–	–	–
	24	49	18	18	–	–
July	1	52	21	15	–	30.76
	8	63	31	14	22.22	28.57
	15	38	22	8	21.05	21.05
	22	–	–	–	–	–
	29	94	26	47	50.00	22.34
August	5	115	55	38	33.04	19.13
	12	60	45	8	13.33	11.66
	19	118	73	24	20.33	17.79
	26	220	149	28	12.72	19.54

Contd...

Table 3–Contd...

Months	Date	No. of Larvae Collected	Moths	Parasitoids	Per cent Parasitism	Mortality Due to Other Factors
September	2	234	89	94	40.17	21.79
	9	304	133	108	35.52	20.72
	16	92	93	43	46.73	28.26
	23	402	197	114	28.35	22.63
	30	399	205	106	26.56	22.05
October	7	404	197	118	29.20	22.02
	14	129	79	38	29.45	9.30
	21	402	125	201	50.00	18.90
	28	121	70	34	28.09	14.04
November	4	524	137	283	54.00	19.84
	11	554	160	282	50.90	20.21
	18	70	32	22	31.42	22.85
	25	501	102	281	56.08	23.55

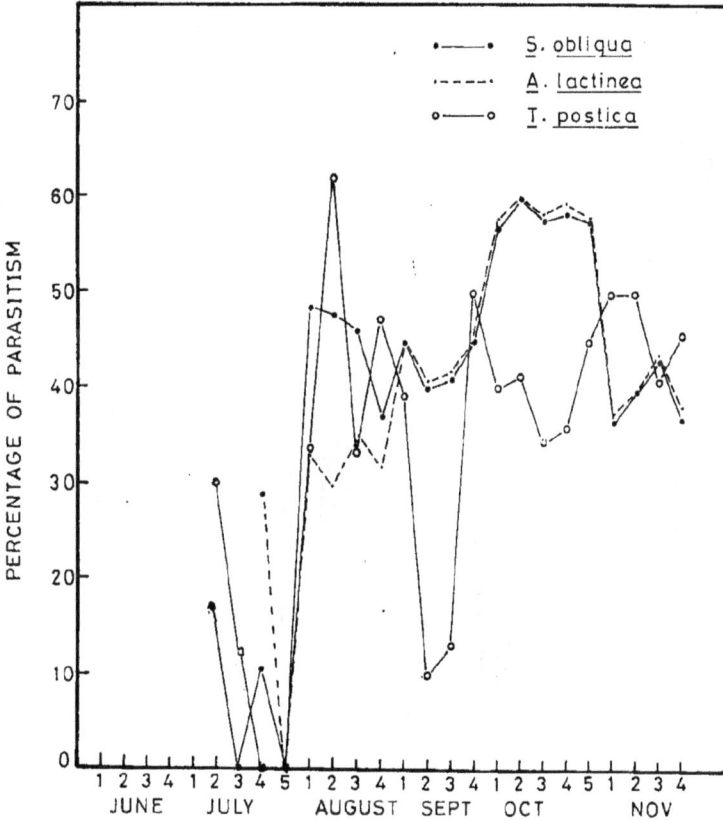

Figure 53: Seasonal Abundance of Parasitoids of
***S. obliqua, A. lactinea* and *T. postica*, 1988**

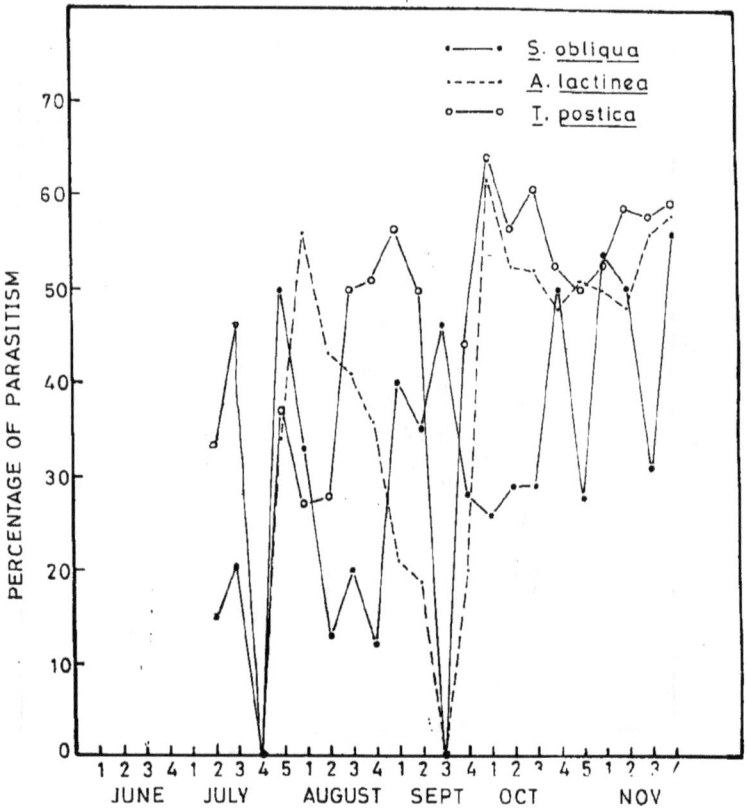

Figure 54: Seasonal Abundance of Parasitoids of
S. obliqua, A. lactinea **and** *T. postica*, **1989**

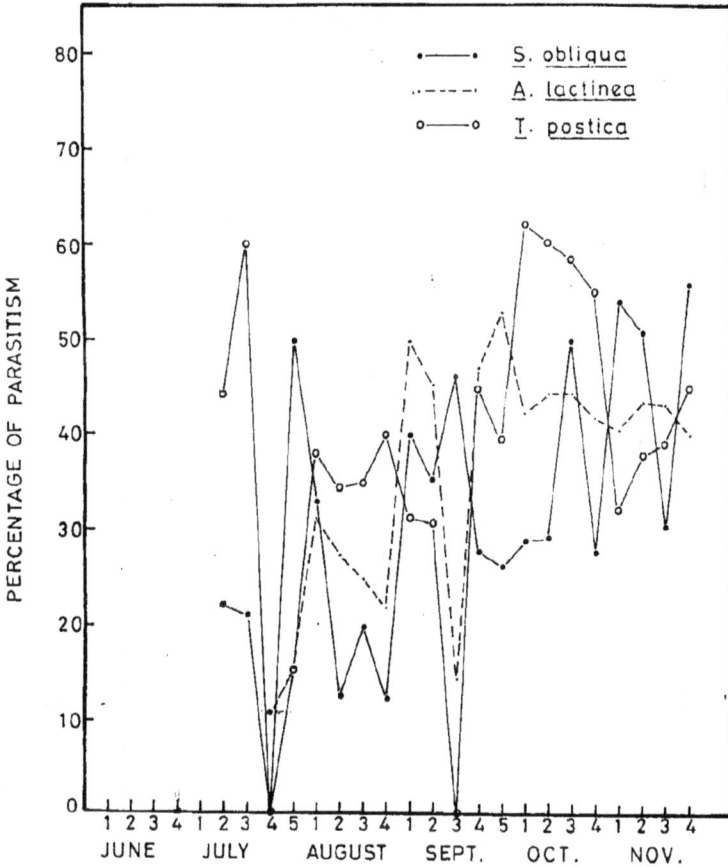

Figure 55: Seasonal Abundance of Parasitoids of
***S. obliqua, A. lactinea* and *T. postica*, 1990**

Table 4: Survey of *A. lactenia* and its Parasitoids on Groundnut (1988)

Months	Date	No. of Larvae Collected	Moths	Parasitoids	Per cent Parasitism	Mortality Due to Other Factors
June	5	–	–	–	–	–
	12	–	–	–	–	–
	19	–	–	–	–	–
	26	–	–	–	–	–
July	3	–	–	–	–	–
	10	12	6	2	–	33.33
	17	14	7	1	–	42.85
	24	48	16	14	29.16	37.50
	31	16	9	–	–	43.75
August	7	78	40	26	33.33	15.38
	14	79	39	24	30.37	20.25
	21	89	9	32	35.95	20.22
	28	103	48	33	32.03	21.35

Contd...

Table 4–Contd...

Months	Date	No. of Larvae Collected	Moths	Parasitoids	Per cent Parasitism	Mortality Due to Other Factors
September	4	146	74	42	28.76	20.54
	11	152	70	48	31.57	22.36
	18	146	77	41	28.08	19.17
	25	172	78	52	30.23	24.41
October	2	172	70	62	36.04	23.25
	9	176	74	61	34.65	23.29
	16	72	31	19	26.38	30.55
	23	90	38	17	18.88	38.88
	30	68	17	21	30.88	44.11
November	6	191	68	71	37.17	27.22
	13	192	62	72	37.50	30.20

Survey terminated

Table 5: Survey of *A. lactinea* and its Parasitoids on Groundnut (1989)

Months	Date	No. of Larvae Collected	Moths	Parasitoids	Per cent Parasitism	Mortality Due to Other Factors
June	4	–	–	–	–	–
	11	–	–	–	–	–
	18	–	–	–	–	–
	25	–	–	–	–	–
July	2	26	5	11	–	–
	9	–	–	–	–	–
	16	8	4	2	–	25.00
	23	–	–	–	–	–
	30	32	9	11	34.37	40.62
August	6	39	7	22	56.41	25.64
	13	48	10	21	43.75	35.41
	20	51	14	21	41.17	31.37
	27	72	25	26	36.11	29.16

Contd...

Table 5–Contd...

Months	Date	No. of Larvae Collected	Moths	Parasitoids	Per cent Parasitism	Mortality Due to Other Factors
September	3	89	48	19	21.34	24.71
	10	100	57	19	19.00	24.00
	17	–	–	–	–	–
	24	118	56	24	20.33	32.20
October	1	200	24	125	62.50	25.50
	6	203	47	106	52.73	23.88
	15	225	50	117	52.00	25.77
	22	237	61	114	48.10	26.16
	29	232	71	119	51.29	18.10
November	5	234	46	117	50.00	30.34
	12	232	51	113	48.70	29.31
	19	229	23	130	56.76	33.18
	26	226	23	131	57.96	31.85

Table 6: Survey of A. lactinea and its Parasitoids on Groundnut (1990)

Months	Date	No. of Larvae Collected	Moths	Parasitoids	Per cent Parasitism	Mortality Due to Other Factors
June	3	–	–	–	–	–
	10	–	–	–	–	–
	17	–	–	–	–	–
	24	–	–	–	–	–
July	1	47	11	22	–	–
	8	12	2	2	–	66.66
	15	8	6	1	–	12.50
	22	9	5	1	11.11	33.33
	29	12	6	2	16.66	33.33
August	5	76	34	24	31.57	23.68
	12	80	36	22	27.5	27.5
	19	82	37	21	25.60	29.26
	26	115	51	26	22.60	33.04

Contd...

Table 6–Contd...

Months	Date	No. of Larvae Collected	Moths	Parasitoids	Per cent Parasitism	Mortality Due to Other Factors
September	2	156	46	78	50.00	20.51
	9	158	54	72	45.56	20.25
	16	14	9	2	14.28	21.42
	23	182	57	86	47.25	21.42
	30	188	58	101	53.72	15.42
October	7	192	67	84	43.75	21.35
	14	182	63	81	44.50	20.87
	21	177	65	79	44.63	18.64
	28	200	63	84	42.00	26.50
November	4	223	78	91	40.80	24.21
	11	205	56	90	43.90	28.78
	18	206	56	89	43.20	29.61
	25	227	71	90	39.64	29.07

T. postica

The survey of *T. postica* was made on ber trees, during June to December. *T. postica* appeared on ber tree in 4th week of June while the parasitoids appeared on 2nd week of July. The population of the pest increased in August and November, decline in population was noticed in September. The percentage of parasitism was highest 62.14 in August. The results on survey of *T. postica* and its parasitoids are recorded in Tables 7–9 and Figures 53–55.

Discussion

Lall (1958), while studying the biology of *Apanteles obliquae* (Walk) reported that the caterpillars of *S. obliqua* were abundant in the field from August to March. Later, Mathur (1962) reported that the pest breaks out twice a year, March to April and August to October. While in present study, the pest was observed from June to December on soyabean. The intensity of population was not found very much decreased during the developmental stage of the crop. However, the population was greatly reduced approaching harvesting period of the crop. Since the pest is polyphagous, reduction in the population may be the result of migration towards other crops. The migration may be due to the maturation of the host plants or harvestation of crops, increase in population density and adverse environmental effects on either the quality of food or ability of the organism to reproduce and survive within the habitat. Migration in *A. lactinea* due to the above regions may be possible but *T. postica* shows specificity towards its host ber, and hence migration factor is less possible than hybernation.

Table 7: Survey of *T. postica* and its Parasitoids on Ber (1988)

Months	Date	No. of Larvae Collected	Moths	Parasitoids	Per cent Parasitism	Mortality Due to Other Factors
June	5	–	–	–	–	–
	12	–	–	–	–	–
	19	–	–	–	–	–
	26	–	–	–	–	–
July	3	89	24	27	–	–
	10	93	23	28	30.12	44.56
	17	18	9	2	11.00	38.88
	24	–	–	–	–	–
	31	–	–	–	–	–
August	7	80	32	27	33.75	26.25
	14	132	16	82	62.12	25.75
	21	148	57	49	33.10	28.37
	28	152	39	72	47.36	26.97

Contd...

Table 7–Contd...

Months	Date	No. of Larvae Collected	Moths	Parasitoids	Per cent Parasitism	Mortality Due to Other Factors
September	4	151	44	59	39.07	31.78
	11	19	9	2	10.52	42.10
	18	29	11	4	13.79	48.27
	25	30	7	15	50.00	26.66
October	2	141	36	57	40.42	34.04
	9	162	52	68	41.97	25.92
	16	172	52	59	34.30	35.46
	23	194	66	71	36.59	29.38
	30	202	59	92	45.54	25.24
November	6	204	49	102	50.00	25.98
	13	208	46	104	50.00	27.88
	20	221	79	91	41.17	23.07
	27	209	61	98	46.88	23.92

Survey terminated

Table 8: Survey of *T. postica* and its Parasitoids on Ber (1989)

Months	Date	No. of Larvae Collected	Moths	Parasitoids	Per cent Parasitism	Unknown Mortality Due to Other Factors
June	4	–	–	–	–	–
	11	–	–	–	–	–
	18	–	–	–	–	–
	25	–	–	–	–	–
July	2	71	28	19	–	33.80
	9	93	34	31	33.33	30.10
	16	101	22	48	47.52	30.69
	23	–	–	–	–	–
	30	27	6	10	37.03	40.74
August	6	37	15	10	27.02	32.43
	13	39	17	11	28.20	28.20
	20	182	32	91	50.00	32.41
	27	181	32	92	50.82	31.49

Contd...

Table 8–Contd...

Months	Date	No. of Larvae Collected	Moths	Parasitoids	Per cent Parasitism	Unknown Mortality Due to Other Factors
September	3	161	22	91	56.52	29.81
	10	182	32	91	50.00	32.41
	24	49	9	22	44.89	36.73
October	1	142	13	91	64.08	26.76
	8	162	43	92	56.79	16.66
	15	163	37	99	60.73	16.56
	22	191	49	101	52.87	21.46
	29	202	44	102	50.49	27.72
November	5	204	38	108	52.94	28.43
	12	208	35	122	58.65	24.51
	19	230	38	131	56.95	26.52
	26	222	22	132	59.45	30.63

Table 9: Survey of *T. postica* and its Parasitoids on Ber (1990)

Months	Date	No. of Larvae Collected	Moths	Parasitoids	Per cent Parasitism	Unknown Mortality Due to Other Factors
June	3	–	–	–	–	–
	10	–	–	–	–	–
	17	–	–	–	–	–
	24	–	–	–	–	–
July	1	47	5	23	–	40.42
	8	49	6	22	44.89	–
	15	51	2	31	60.78	35.29
	22	–	–	–	–	–
	29	18	7	3	16.66	44.44
August	5	72	11	28	38.88	45.83
	12	79	21	27	34.17	39.24
	19	88	34	31	35.22	26.13
	16	101	27	41	40.59	32.67

Contd...

Table 9–Contd...

Months	Date	No. of Larvae Collected	Moths	Parasitoids	Per cent Parasitism	Unknown Mortality Due to Other Factors
September	2	102	22	32	31.37	47.05
	9	100	28	31	31.00	41.00
	16	–	–	–	–	–
	23	17	3	8	47.05	35.29
	30	28	6	11	39.28	39.28
October	7	129	7	81	62.79	31.78
	14	130	21	78	60.00	23.84
	21	135	35	79	58.51	15.55
	28	136	40	76	55.88	14.70
November	4	155	53	51	32.90	32.90
	11	159	49	62	38.99	30.18
	18	178	58	71	39.88	27.52
	25	202	54	92	45.54	27.42

In *Helicoverpa armigera* Hubner, Bilapate *et al.* (1979) studied the population dynamics on sorghum, pigeonpea and chickpea in Marathwada region of Maharashtra state. Their observations revealed that on jowar, the larval population in the 1st and 2nd instars was reduced to 3.33 per cent by a larval parasitoid. In pigeonpea, the most frequently encountered pupal parasitoid was a *Tachinid* fly. The population of pest was reduced in the pupal stage to the extent of 50 per cent on pigeonpea during October, 1977. Parasitization was less (3.73 per cent) in early instars 1st and 2nd) and increased (24.92 per cent) thereafter 3rd and 4th instars. The parasitoid noticed by them was *C. chlorideae* which contributed to a reduction in the population of the pest in 3rd and 4th instars. However, the incidence of this parasitoid declined (0.36 per cent) in the late instars (5th and 6th).

Faleiro *et al.* (1986) studied the pest complex population dynamics on cowpea, wherein they noted that the regular insect pests occurring during Kharif season were the green jassid *Amrasca kerri* Pruthi; semilooper *Plusia orichalcea* F.; white fly *Bemisia tabaci* Genn; leafminer *Acrocercops phaeaspora* Meyr.; coreid bug *Cletus* sp; green stink bug *Nezara viridula* (L.); flower beetle *Mylabris pustulata* (Thnb.) and the green pod borer *Maruca testulalis* (Geyer). The green stink bug and the green pod borer were observed during Kharif 1983 and Kharif 1984 respectively. The sporadic insect pests occurring during the Kharif season were the galerucid beetle *Madurasia obscurella* Jac; the surface grasshopper *Chrotogonus trachypterus* (Blanch.); Bihar hairy caterpillar *Spilosoma obliqua* (Wlk.); green slender bug *Creontiades pallidifer* (Wlk.) and thrip (unidentified). The stray insects occurring during Kharif were aphid

Aphis craccivora Kach.; grasshopper *Oxya velox* F.; grey weevil *Myllocerus undesimpstulatus* Desb.; black beetle *Cyrtozemia cognata* Marsh, painted bug *Bagrada hilaris* (Burm.); red pumpkin beetle *Raphidopalpa foveicollis* (Lucas.); red cotton bug *Dysdercus koenigii* (F.); podweevil *Apion* sp. and the pulse beetle *Callobruchus chinensis* (L.). The black beetle, painted bug red pumpkin beetle and red cotton bug occurred only during Kharif 1983.

Studies on the trapping of adult males of *Phthorimaea operculella* (Zeller) on sex pheromone baited water trap was conducted by Raj (1988) at Central Potato Research Station, Rajgurunagar (Pune) from 1981 to 1983. His observations revealed that moths were on wing throughout the year in Deccan plateau and there were no hibernation or aestivation of this pest in this area. There were 2 main peaks in its population buildup; 1st in 9th week (early March) in rabbi crop and 2nd in 43rd week (October) after Kharif crop. Besides there were 4 smaller peak in 12th week (March), 16th week (April), 30th week (July) and 39th week (September). A number of mini peaks were also present. The pheromone trapping appeared to be very useful for diagnostic investigation.

Investigations on sampling for estimation of larval population of *H. armigera* on 2 varieties of chickpea carried out (Kaushik and Naresh, 1989) at Haryana Agricultural University revealed that Ground Cloth-shake Method (GCSM) was more efficient, accurate and practical than Visual Count Method (VCM). Sample unit size of 1:0 m row length was more precise than 0.5 m and more economical than 1:5 m. Eight samples of 1.0 m row length per lot (48 m²) were sufficient to obtain 90 per cent accuracy in sampling. The study survey of the species

S. obliqua and *A. lactinea* was carried out by keeping time constant (1 hr) for search and survey was terminated after the crop harvestation.

Singh *et al.* (1989) studied the population dynamics of Mustard aphid, *Lipaphis erysimi* Kalt. Their results indicated that the aphid was found to appear and establish on *Brassica* spp. in the third week of December. It built up its population in January–February reaching the peak on 8th and 18th February in 1980 and 1981, respectively. The environmental parameters played an important role in the aphid infestation. The ambient maximum (21.68–23.52°C) and minimum (7.18–9.40°C) temperatures in January–February appeared to be most conducive for the aphid multiplication. High humidity had little impact on aphid population fluctuation. Minimum humidity, ranging from 55.7–69.4 per cent in January–February favoured the population build up while the activity of the aphid ceased at 59.90 per cent and below. Frequent rains during the population rise phase (January–February) adversely affected the aphid population. *T. postica* population increased twice, in August and November in Kolhapur region.

Goyal and Kumar (1989) reported the seasonal abundance of insect pests on monsoon crop, sunflower; their phenological studies revealed a vital impact of the minimum temperature and relative humidity for insect abundance in a monsoon crop of sunflower (EC 68414). Significantly high coefficient of correlation have been measured to assess their role during the successive growth of crop. Insect species dominance were also discussed because of their preferential behaviour. Frequent as major insect pests, the first order consumers were parasitized and preyed upon by certain insect species listed second order consumers. Species of major numerical dominance associated with different parts of the crop were *Attractomorpha crenulata* (F.), *Leptocentrus taurus* (F.), *Myllocerus lateralis* Chev., *Sarcopleaga* sp., *Carpophilus* sp., *Brumus suturalis* (F.), *Camponotus compressus* (F.), and several lepidopteran caterpillars. The average agro-climatic factors of the cropping period were 34.6±0.22°C max. and 458.5 mm rainfall during June to August, 1979.

Standard Period and Weeks

Period No.	Week No.	Dated
I	1 January	1–7 January
	2 January	8–14
	3 January	15–21
	4 January	22–28
	5 January	29–4 February
II	6 February	5–11
	7 February	12–18
	8 February	19–25
	9 February	26–4 March*
III	10 March	5–11
	11 March	12–18
	12 March	19–25
	13 March	26–1 April
IV	14 April	2–8
	15 April	9–15
	16 April	16–22
	17 April	23–29
	18 April	30–6 May
V	19 May	7–13
	20 May	14–20
	21 May	21–27
	22 May	28–3 June

Period No.	Week No.	Dated
VI	23 June	4–10
	24 June	11–17
	25 June	18–24
	26 June	25–1 July
VII	27 July	2–8
	28 July	9–15
	29 July	16–22
	30 July	23–29
	31 July	30–5 August
VIII	32 August	6–12
	33 August	13–19
	34 August	20–26
	35 August	27–2 September
IX	36 September	3–9
	37 September	10–16
	38 September	17–23
	39 September	24–30
X	40 October	1–7
	41 October	8–14
	42 October	15–21
	43 October	22–28
	44 October	29–4 November
XI	45 November	5–11
	46 November	12–18
	47 November	19–25
	48 November	26–2 December
XII	49 December	3–9
	50 December	10–16
	51 December	17–23
	52 December	24–31**

*: In leap years the last week of period I will be 26 February to 4 March *i.e.* 8 days instead of 7.

**: Last week of period XII have 8 days, 24–31 December.

Chapter 4

Species Diversity

Introduction

Individuals of a species differ from one another inherently in their genetic properties and also as a result of environmental influences on them during their development. Most species represent a considerable spectrum of genetic variability. The existence of these varieties implies some selection for adaptation, not only to climate and cultural conditions, but to pest as well. Mapping out the diversity of the insect world is in many ways anologues to the work of the topographer. The amount of diversity in the world is staggering. About 1 million species of insects already been described and estimates on the number of many more are still undescribed. Each species may exist in numerous different forms like sexes, age classes, seasonal forms, morphs and other phena. If not ordered and classified, it would be impossible to deal with this enormous diversity. For ordered the rich diversity of the animal world and to develop methods and principles to make this task possible, systematic Zoology endeavours. Keeping in view the importance of diversity of insect pests specially, hairy caterpillars of economic agricultural crops, the present chapter has been visualized.

Materials and Methods

Various types of hairy caterpillars were collected from different crops in Maharashtra State. The caterpillars collected from different parts have been reared under laboratory conditions (25±1°C, 55–60 per cent RH, 12 hr. photoperiod). Later, these species were identified. Adult forms of the laboratory reared caterpillars are kept in collection of T.V. Sathe for time being and will be deposited to Zoological Survey of India, Calcutta.

Results

During the extension survey of hairy caterpillars from June 1988 to November 1990, a large number of hairy caterpillars have been collected from different fields of crops.

The results on species diversity of hairy caterpillars of economic important crops tabulated in Table 10, indicated that the most abundant-hairy caterpillars were *S. obliqua, A. moorei, A. lactinea.* Scorpion caterpillar on pulses and oilseed crops. The caterpillars *T. visnu* and *T. postica* were found on castor and ber respectively. Scorpion, caterpillars found attacking the groundnut from August to September with abundance. It was also found on Pigeonpea from October to January. The seasonal abundance of caterpillars, nature of damage and the crops which they attacked are shown in Table 10 and Plate II, Figures 8–18; Plate III, Figures 19–29.

Table 10: List of Some Hairy Caterpillars on Some Crops of Economic Importance

Sl.No.	Common Name	Scientific Name	Order	Family	Crop	Nature of Damage	Seasonal Abundance
1.	Red hairy caterpillar	Amascta moorei (Butler)	Lepidoptera	Arctiidae	Groundnut, sunflower, castor, maize, sorghum	Defoliation	July, November
2.	Red hairy caterpillar	A. albistriga (Walker)	Lepidoptera	Arctiidae	Groundnut, pearl millet, cowpea, castor, cotton etc.	Defoliation	July, August
3.	Hairy caterpillar	A. lactinea (Cramer)	Lepidoptera	Arctiidae	Groundnut	Defoliation	August, September
4.	Bihar hairy caterpillar	Spilosoma obliqua (Walker)	Lepidoptera	Arctiidae	Pulses, pea, cotton etc.	Defoliation	March, April, August, October
5.	Castor hairy caterpillar	Euproctis lunata (Walker)	Lepidoptera	Lymantriidae	Groundnut, linseed, castor	Defoliation	July, September
6.	Hairy caterpillar	E. guttata (Walker)	Lepidoptera	Lymantriidae	Castor	Defoliation	February
7.		E. scientillans (Walker)	Lepidoptera	Lymantriidae	Castor	Defoliation	December

Contd...

Table 10–Contd...

Sl.No.	Common Name	Scientific Name	Order	Family	Crop	Nature of Damage	Seasonal Abundance
8.	Castor spiny caterpillar	*Ergolis merione* (Cramer)	Lepidoptera	Nymphalidae	Castor	Defoliation	August
9.	Hairy caterpillar	*Trabala vishnu* (Lefebure)	Lepidoptera	Lasiocampidae	Castor	Defoliation	March
10.	Ber hairy caterpillar	*Thiacidas postica* (Walker)	Lepidoptera	Arctiidae	Ber	Defoliation	August, November
11.	Scorpion caterpillar	*Euproctis traterna*	Lepidoptera	Lymantridae	Groundnut, pigeonpea	Defoliation	August, September, October, January
12.	Hairy caterpillar	*Pericallia ricini* (Fabr.)	Lepidoptera	Arctiidae	Castor, cucurbit	Defoliation	October, November
13.	Syntomid caterpillar	*Syntomis cyssea* (Cramer)	Lepidoptera	Syntomidae	Bean	Defoliation	July, August
14.	Pyraustid caterpillar	*Phlyctaenia tyres* (Cramer)	Lepidoptera	Pyraustidae	Bean	Defoliation	July, August

Discussion

Hairy caterpillars attacked number of agricultural crops and caused serious damage which resulted in considerable reduction in the yield of crops. Some of the hairy caterpillars like *S. obliqua, Amsacta* spp., *Eproctis* spp. are polyphagous in their habit and hence they appeared cronically and alternatively on the crops.

According to Gupta *et al.* (1966) *A. moorei* is widely distributed throughout India and the caterpillars are highly polyphagous which defoliate number of Kharif crops. Gupta *et al.* (1966) also reported that *A. moorei* attack groundnut, sunflower, castor, maize, sorghum, pearl-millet (*Pennisetum typhoides*), dew-gram (*Phaseolus aconitifolius*). lablab (*Dolichos lablab*), black-gram (*Phaseolus mungo*), sannhemp (*Crotalaria juncea*), soyabean (*Glycine max*), cluster-bean (*Cyamopsis tetragonoloba*), green-gram (*Phaseolus aureus*), rice (*Oryza sativa*), jute (*Corchorus* spp.), sweet-pototo (*Ipomoea batatas*), kodo-millet (*Paspalum scrobiculatum*), lucerne (*Medicago sativa*), etc. Patel and Patel (1965) recorded this pest on tobacco (*Nicotiana* spp.), pigeonpea (*Cajanus cajan*), cotton (*Gossypium* spp.), *Opuntia* sp. while, Bindra and Kittur (1961) and Srivastava and Goel (1962) reported the same pest on Cowpea (*Vigna sinensis*) and some fruit trees respectively.

A. albistriga is rarely found in Kolhapur region. However, it is a serious polyphagous pest in most of the red scandy lom tracks of Southern India and infest rainfed crops. It damages specially ground-nut and pearl-millet. The caterpillars generally, appear in millions and move from field to field in a particular direction. Ali (1961) studied the population dynamics of this pest in field condition where he estimated 5.5 million caterpillar's population per hectare.

The detailed biology of the species was studied by Nagarajan *et al.* (1957).

A. lactinea is reported on groundnut and pulses like cowpea and pigeonpea in Maharashtra. Sen and Mukharjee (1955) reported the species attacking groundnut in West Bengal and Srivastava *et al.* (1965) reported that *A. lactinea* is often accompanied with *A. moorei* in Uttar Pradesh.

S. obliqua is very important oilseed crop pest in Maharashtra and most of the sunflower fields of Marathawada and Southern Maharashtra are badly damaged. In Maharashtra, it is also reported on variety of crops like groundnut, cow pea, pigeonpea, castor, maize, linseed, etc. Lefroy (1907a) and Srivastava (1964) reported the species on groundnut, sunflower, sesmum linseed, jute, sunhemp, cotton, rice, pea, lucerne, pearl-millet, pulses and maize.

Euproctis spp.

Three species of *Euproctis* (Table 10) are recorded from different parts of India. *E. lunata* is reported from Delhi (Anonymous, 1953) and Rajasthan (Kushwaha and Bhardwaj, 1967). It is sporadic pest of castor (Pandey, 1968). The caterpillars defoliate the crop. Narayanan (1959) reported that the pest is active throughout the year and several generations are seen in a year. But, Kushwaha and Bhardwaj (1967) reported only 3 generations in a year. *E. guttata* is recorded from Hyderabad attacking leaves and capsules of the castor. Khan and Rao (1948) collected the caterpillars from *Bauhinia* spp., *Ziziphus mauritiana*. Khan and Rao (1948) also reported *E. scintillans* at Hyderabad. They noted the larval feeding gregariously on castor leaves. The above three species are also reported from Maharashtra.

LONGEVITY, NUTRITION
AND DEVELOPMENT

Introduction

Nutrition is very important abiotic population regulatory factor of insects in nature and laboratory. The ability of the organism depends on the nutritional requirements, hence there exist a direct and essential connection between an environmental factors and vital processes of the insect. However, the concept of use of nutritional requirements in insect control is not yet fully studied and hence, not applied properly in the pest management programmes. The failure of recognize a suitable food for the insects, resulted many times unsuccessful attempts in pest management programme. The effectiveness of control measures entirely depends on quality and nature of food specially in biological control programme. Keeping in view the above facts, different dietary combinations have been tried against the immature forms and the adult pest species.

Materials and Methods

Effect of nutritional requirements of various food plants have been studied by providing young leaves of groundnut (*Arachis hypogea* L.), Castor (*Ricinus communis* L.), Sunflower (*Helianthus annus* L.), Cowpea (*Vigna unguculata* L.), Soyabean (*Glycine max* L.) and ber (*Ziziphus mauritiana* L.). Newly emerged 100 caterpillars of *S. obliqua* and *A. lactinea* and *T. postica* were fed with above maintained food plant leaves and observations were made on larval length, pupal length and percentage of adult emergence. Fresh leaves were provided to larvae at 12 hr interval.

To observe the effect of feeding on longevity of adults, the experiments were carried out by feeding of moths on water alone, 10 per cent honey, 10 per cent sucrose, 10 per cent glucose D and protinex 5 per cent + sugar 5 per cent. In control, adults were starved.

The experiments were also carried out to see the development of species on artificial diet. The diet consisted:

1. Host food plant seed containt :   25.00 g
2. Ascorbic acid   :   3.50 g
3. Sorbic acid   :   1.00 g
4. Dry leaves powder   :   75.00 g
5. Sucrose   :   5.00 g
6. Dist. water   :   100.00 ml

The artificial diet was tried against *S. obliqua*, *A. lactinea* and *T. postica* at laboratory conditions (25±1°C, 55–60 per cent RH, 12 hr photoperiod. The artificial diet was provided to the caterpillars in petridishes and larval and pupal duration was counted. Observations were also made on percentage emergence of adults. All above experiments were replicated for five times for confirming the results.

Results

S. obliqua

Larval and pupal length was studied by providing castor, sunflower, groundnut, cowpea, soyabean and ber leaves to the caterpillars of *S. obliqua*. The results showed that castor leaves were most suitable for fast development of this pest. Sunflower leaves ranked IInd, while ber leaves were unsuitbale for the development. The results are recorded in Table 11.

The larval and pupal development was completed within shortest period. 30.50 days when fed with castor leaves, while it was prolonged for 38.00 days when fed with cowpea leaves. The percentage of adult emergence was also highest with castor leaves (Table 11). The adult longevity was highest, 9.6 (range 8–10) days in females and lowest 5.2 (range 4–6) days in males (Table 12). The artificial diet tested was suitable for the larval and pupal development but its per cent adult emergence was less, 68 per cent (Table 17).

A. lactinea

Same food plants were tried against *A. lactinea* and observations were made on larval and pupal length of the species. The results showed that groundnut leaves were most suitable for fast development of *A. lactinea* while cowpea were least (Table 13). The order of preference on the basis of fast development was as groundnut > castor > sunflower > soyabean > cowpea. The ber leaves were not suitable for rearing the larvae. The adults survived longest, 12 days with 10 per cent honey; other foods 10 per cent sucrose, 10 per cent glucose D, and protinex 5 per cent + sugar 5 per cent have also showed their potential as an important food for this species (Table 14). The artificial diet yielded 65 per cent adults with average developmental period of 40 days (Table 17).

T. positca

The results on food plant preference of *T. postica* are represented in Table 15. The results indicated that the larvae of *T. postica* prefered only the leaves of ber most and castor

Table 11: Effect of Food Plants on Food Preference and the Development of *S. obliqua*

Sl.No.	Food Plants	No. of Larvae Tried	Total Larval Duration (Days)	Total Pupal Duration (Days)	Per cent of Adult Emergence
1.	*Ricinus communis* L.	100	20.00	10.50	100 per cent
2.	*Arachis hypogaea* L.	100	23.50	13.00	98 per cent
3.	*Helianthus annuus* L.	100	21.50	11.50	99 per cent
4.	*Glycine max.* L	100	22.00	12.00	98 per cent
5.	*Vigna uniquiculata* (L.)	100	24.50	13.50	98 per cent
6.	*Ziziphus mauritiana* L.	100	0.00	0.00	00

Table 12: Effect of Different Food on *S. obliqua* Adult Longevity and Pre-oviposition Period

Sl.No.	Treatments	Preoviposition Period	Longevity of Adults (Days)	
			Male	Female
1.	10 per cent honey	1.50	9.4 (8–10)	9.6 (8–10)
2.	10 per cent sucrose	2.50	8.0 (7–9)	8.2 (7–9)
3.	10 per cent Glucose D	2.50	8.4 (7–9)	8.4 (7–9)
4.	Protinex* 5 per cent + sugar 5 per cent	2.00	8.6 (7–9)	8.6 (7–9)
5.	Water	3.00	6.0 (5–7)	6.2 (5–7)
6.	Control	4.00	5.2 (4–6)	5.4 (4–6)

*: Protinex is a product of M/s Pfizer Limited, Bombay and contains carbohydrates, protein, vitamin and minerals.

* Figures in parenthesis are ranges.

Table 13: Effect of Food Plants on Food Preference and the Development of *A. lactinea*

Sl.No.	Food Plants	No. of Larvae Tried	Total Larval Duration (Days)	Total Pupal Duration (Days)	Per cent of Adult Emergence
1.	R. communis	100	24.00	12.5	100 per cent
2.	A. hypogaea	100	23.00	12.00	100 per cent
3.	H. annuus	100	24.5	13.00	100 per cent
4.	G. max	100	25.00	13.5	98 per cent
5.	V. uniquiculata	100	26.50	13.5	96 per cent
6.	Z. mauritiana	100	00.00	00.00	0.00 per cent

Table 14: Effect of Differen Food on *A. lactinea* Adult Longevity and Pre-oviposition Period

Sl.No.	Treatments	Preoviposition Period	Longevity of Adults (Days)	
			Male	Female
1.	10 per cent honey	1.5	10.4 (9–12)	10.6 (9–12)
2.	10 per cent sucrose	2.5	8.2 (7–9)	8.2 (7–9)
3.	10 per cent glucose D	2.0	9.00 (8–11)	9.2 (8–11)
4.	Protinex* 5 per cent + sugar 5 per cent	2.0	10.2 (8–11)	10.4 (8–11)
5.	Water	3.0	4.0 (4–5)	4.0 (4–5)
6.	Control	3.5	3.6 (2–5)	3.8 (2–5)

*: Protinex is a product of M/s Pfizer Limited, Bombay and contains carbohydrates, protein, vitamin and minerals.

*: Figures in parenthesis are ranges.

Table 15: Effect of Food Plants on Food Preference and the Development of *T. postica*

Sl.No.	Food Plants	No. ol Larvae Tried	Total Larval Duration (Days)	Total Pupal Duration (Days)	Per cent of Adult Emergence
1.	R. communis	100	23.5	12.2	60.00
2.	A. hypogaea	100	–	–	–
3.	H. annuus	100	–	–	–
4.	G. max	100	–	–	–
5.	V. unquiculata	100	–	–	–
6.	Z. mauritiana	100	15.5	8.00	100 per cent

Table 16: Effect of Different Food on *T. postica* Adult Longevity and Pre-oviposition Period

Sl.No.	Treatments	Preoviposition Period	Longevity of Adults (Days)	
			Male	Female
1.	10 per cent honey	1.00	9.2 (8–12)	9.2 (8–12)
2.	10 per cent sucrose	2.50	7.4 (6–10)	7.4 (6–10)
3.	10 per cent glucose D	2.00	7.8 (7–11)	7.8 (7–11)
4.	Protinex* 5 per cent + sugar 5 per cent	1.50	8.2 (7–11)	8.4 (7–11)
5.	Water	3.00	5.0 (5–6)	5.2 (5–6)
6.	Control	3.00	3.6 (3–4)	3.8 (3–4)

*: Protinex is a product of M/s Pfizer Limited, Bombay and contains carbohydrates, protein, vitamin and minerals.

*: Figures in parenthesis are ranges.

least. Other food plants tried were not suitable for the development of this species. The adults survived maximum, 12 days with an average period of 9.2 days. If starved the adults died within 3–4 days (Table 16). Artificial diet yielded 69.0 per cent adults (Table 17) and thus proved its potential, but it was not optimum food for the *T. postica* rearing.

Discussion

Nutrition developmental interactions in some of the insect pests have been studied by Thobbi (1961), Mathur (1962), Pandey and Srivastava (1967), Manoharan *et al.* (1984), Masoodi (1985), Masoodi and Srivastava (1985), Poonia (1985), Chandra *et al.* (1985), Sharma and Chaudhary (1985), Anand and Pant (1986), Singh and Mavi (1986), Singh and Ram (1987), Devaraj and Subramanya (1987), Jayanathi and Metha (1987), Madan *et al.* (1987, 1988), Singh *et al.* (1987), Srivastava *et al.* (1987), Goyal and Rathore (1988), Tiwari *et al.* (1988), Dakshayani *et al.* (1988), Bhalani (1989), Bhatia and Sethi (1989), Paripurna and Srivastava (1989), Srivastava and Pant (1989), Methew *et al.* (1990) etc. The review of literature showed that the food plant spectrum of *S. obliqua* have been studied previously by some workers. However, some food plants in the present study have been tested and the food spectrum of *A. lactinea* and *T. postica* reported for the first time.

Manoharan *et al.* (1984) studied the influence of food quality on food preference and the development in *E. fraterna. R. communis* was found to be a highly suitable food plant for *E. fraterna* with reference to rapid growth and high fecundity. *S. melongena* was the least preferred among the food plants tested. The results indicated that

Table 17: Effect of Artificial Diet on Development and Survival

Sl.No.	Species	No. of Larvae Tried	Total Larval Duration (Days)	Total Pupal Duration (Days)	Per cent of Adult Emergence
1.	S. obliqua	100	25.50	13.00	68.00
2.	A. lactinea	100	27.00	13.00	65.00
3.	T. postica	100	23.00	12.00	69.00

the growth of the larval instars and fecundity of *E. fraterna* were positively correlated with the concentration of protein and water and negatively correlated with the concentration of ash in the leaves of food plants used. In present study castor leaves, groundnut leaves and ber leaves proved to be the best food material for the development of *S. obliqua, A. lactinea* and *T. postica* on which the larval and pupal developments were completed within 30.5. 35 and 23.5 days respectively.

Masoodi and Srivastava (1985) studied the effect of host plant on pupal weight and fecundity of *Lymantria obfuscata* Walker (Lepidoptera). In all fifteen host plants including four varieties of apple (red delicious, americon apirouge, benani and ambri), two varieties of pear (william pear and sand pear), quince, two species of popular (*Populus alba* and *P. nigra*), almond, cherry, apricot and walnut were tested for their influence on the pupal weight and fecundity of *L. obfuscata*. The weight of male and female pupae and the fecundity of the females, developed on leaves of red delicious, american apiroug *S. fragilis* and *P. nigra*, was higher than those developed on *P. alba*, apricot, cherry, almond and walnut. Reproductive index also suggested greater efficiency on diets producing female pupae of higher weight.

Masoodi (1985) reported growth response of *L. obfuscata* in relation to tannin content in fifteen host plants. He noted that the tannin content of foliage increased with the maturation of leaves. Sand Pear and William Pear contained higher tannin content and the larvae of *L. obfuscata* did not survive on their foliage. Apricot, walnut, cherry and almond had high tannin contain initially which was not found conducive to the growth of the pest. The development of larvae on these host plants was slower with reduced pupal weight. The survival percentage was low. Larval development was rapid with higher survival and pupal weight on three varieties of apple (Red Delicious, American Apirouge and Ambri) *S. alba* and *P. nigra.* The foliage of these host plants contained low tannin content ranging between 0.33 and 0.83 per cent of dry weight. Quince *S. babylonica, P. alba* and Benani variety of apple did not vary, much in the tannin content of their foliage and development was slower than other varieties of apples, poplar and willow. The results suggested a direct relationship between the tannin content of host plant and development and survival of *L. obfuscata.*

Sharma and Chaudhary (1985) studied the reproductive behaviour of *Helicoverpa armigera* (Hubner) with respect to the adult nutrition, they noted that feeding of 10 per cent sugar was most effective in reducing pre-oviposition period (5.2 days), increasing oviposition period (20.2 days), fecundity per female (1501.2 eggs), fertility (16.4 per cent) and longevity of males (2.0 days) and females (25.2 days). The next best food was mixture of 5 per cent protinex + 5 per cent sugar solution as in this case, preoviposition and oviposition periods were 5.6 days and 14.2 days, respectively and the fecundity per female, fertility and longevity of males and females were 1302.0 eggs, 69.7 per cent, 28 ad 20 days, respectively. The fecundity (64.8 eggs), fertility (61.2 per cent) and longevity of males (11 days), and females (13.2 days) was very much reduced among the moths fed on 10 per cent protinex solution, indicating the moths of *H. armigera* were unable to assimilate protein. Also only 20 per cent females mated when fed on this diet and the oviposition period (8 days) was very much reduced even in those moths which laid eggs. In case of those fed on water alone, the mating could take place only in 40 per cent females and the pre-oviposition period (3 days), oviposition periods (4 days), fecundity (10.4 egg/female), fertility (59.6 per cent) and longevity of males (8.4 days) and females (9.6 days) was exceptionally poor. The moths died without any feeding and without egg laying within 5-9 days.

Anand and Pant (1986) studied the effect of various vitamins of B-complex on the growth and survival of *Chilo partellus* (Swin.) with the help of chemically defined diet by delation technique. Their results revealed that in the absence of riboflavin, biotin or all vitamins, there was no pupation at all and in the absence of folic acid only 5 per cent pupation was recorded. These three vitamins seems to be absolutely essential. When thiamine or nicotinic acid as omitted from the diets, the pupal percentage fall down to 30 or 40 per cent respectively. This reduction was statistically significant compared to that in control. In other diets lacking pantothenic acid, pyridoxin. p-amino benzoic acid, inositol or choline chloride pupal percentage was as good as in diet with all the vitamins. They have concluded that riboflavin, biotin and folic acid were absolutely essential in the diet of *Chilo* larvae and importance of thiamine and nicotinic acid was also shown to some extent, while rest of the other vitamins were not required by this insect.

Singh and Mavi (1986) observed the growth and development of leaf miner, *Phytomyza horticola* Goureau with respect to some *Brassica* hosts. They reported that larvae and pupae of *P. horticola* reared on different host measured significantly more on BSH-1 than on *B. tournefortii* and RLM 198. Non-significant variations in the weight of males was found when cultured on different hosts but weight of females varied significantly during the first 3 generations, being more on BSH-1. Total body length of adults was significantly more when reared on BSH-1 than that on *B. tournefortii* and RLM 198. Thus, BSH1 was the more suitable host for the development of *P. horticola*.

Devraj and Subramanya (1987) studied the effect of antibiotics on growth, development and life cycle of *Spodoptera litura* Fab. They reported that in *S. litura* reared on natural food (castor leaves) and semisynthetic food, antibiotics administered through natural diet, Chloramphenicol, tetracycline and chlortetracycline supplemented with semisynthetic diet markedly reduced pupation while, the above three antibiotics as well as oxytetracycline produced similar effects when administered through castor leaves. There was significant reduction in the mean growth index with all the antibiotics supplemented with either of the diets. Tetracycline in artificial diet markedly reduced the mean pupal weight while, none of the antibiotics had any effect when fed through castor leaves. Chlortetracycline markedly increased the weight of the adults when administered through artificial diet while, oxytetracycline and tetracycline lowered the adult weight when fed through castor leaves. In the present study artificial diet was prepared with seed content, dry leaves powder of host food plant, sucrose, ascorbic acid and water (proportion is given in methodology) and fed to *S. obliqua, A. lactinea* and *T. postica* but not found optimum.

The castor semilooper, *Achaea janata* Linn. was reared by Singh and Reddy (1987) on a oligidic diet containing castor leaf powder, corn germ and castor oil. Their results showed that the insect's requirement for castor leaf powder and corn germ was 12 and 5 per cent respectively. The principal results showed that dietary supplement of caesin (3.5 per cent), carbohydrate (2.5 per cent) sterols (0.2 per cent), mineral mixture (1 per cent) and L-ascorbic acid (0.5 per cent) were essential for optimum growth and development of the insect.

The experiments were conducted by Tiwari *et al.* (1988) to study the growth and developmental behaviour of *S. obliqua* on nine varieties of groundnut namely, AH-1192, C-501, Dwarf Mutant, J-11, JH-62, M-13, M-145, OG-71-3 and PO1-2. Their observations were on larval period, prepupal period, pupal period, pupal weight, percent pupation, per cent adult emergence, adult longevity, fecundity and per cent egg hatching. Prolongation of larval period was recorded on M-13 while, faster development was observed on M-145. Longest pupal period was recorded on M-145 while, on J-11 and JH-62 pupal period was shortened considerably. Pupae obtained on M-145 and C-501 gained higher weight. Considerably high population was recorded on M-145 and Dwarf Mutant. The adult longevity did not show significant difference on test varieties. Highest growth index was observed on M-145 while, it was lowest on M-13. In the present study leaves of castor, groundnut, sunflower, cowpea, soyabean and ber were provided and developmental periods in *S. obliqua, A. lactinea* and *T. postica* were recorded. Castor leaves were most suitable for *S. obliqua*, groundnut leaves for *A. lactinea* and ber leaves for *T. postica* on which the larval-pupal development was completed in 30.5, 35 and 23.5 days respectively. However, *S. obliqua* has been reported to feed on sugarbeet *Beta vulgaris* Linneaus in India by Lefroy

(1907), Lall (1964), Pruti (1969) and Chhibber (1975). Studies on the growth and development of this insect have earlier been carried out on several field crops (Pandey *et al.*, 1968; Katiyar *et al.*, 1975; Deshmukh *et al.*, 1977; Yadav *et al.*, 1878), weeds (Rethore and Sachan, 1978), Medicinal plants (Lal and Mukharji, 1978) and ornamental plants (Rathore and Sachan, 1981).

Goyal and Rathore (1988) studied the susceptibility of different hosts to *H. armigera*. They reported that the susceptibility of the host plants to the pest was in decreasing order, as gram < pea < linseed < arhar < tomato < cotton. While, in *T. postica* the order of preference of host plants was toward ber castor. The order of preference in *A. lactinea* was as groundnut > castor > sunflower > soyabean > cowpea > while, in *S. obliqua* the order of preference was as castor > sunflower > soyabean > groundnut > cowpea.

Attempts were made by Dakshayani *et al.* (1988) to rear the rice leaf folder *Cnaphalocrosis medinalis* larvae on artificial diet. Their results showed that none of the test diets except two (*C. partellus* and *C. polychrysa* diets) were suitable for rearing *C. medinalis* larvae. On rest of the diets, nenonate larvae did not survived beyond 5 days while, second instar larvae survived for 5 to 10 days though none pupated. On the other hand, on *C. partellus* diet supplemented with rice leaf (15 and 30 per cent by wt), 5 to 6 per cent second instar larvae completed development and pupated on *C. polychrysa* diet in original composition (Kalode *et al.*, 1970) 3 per cent of the larvae completed development. In general, rice leaf powder was found to be an essential component as diets, lacking it showed poor growth.

Bhalani (1989) reported the suitability of host plants for growth and development of leaf eating caterpillar

S. litura. His results showed that the larvae of *S. litura* were reared on 7 natural food plants and their effect on the larval and post larval development of the insect was recorded. On the basis of survival percentage of larvae, growth index value; pupal weight, size and period, percentage of adult emergence and fecundity of moths, castor was the most suitable host for *S. litura.* Maize was the least preferred of the food plants tested and was found to have a distinct retarding effect on the growth rate of *S. litura* resulting in prolonged larval period, lowest survival percentage, minimum pupal weight and growth index values. On the basis of the growth index values the remaining food plants could be arranged in the following descending order: cotton > groundnut > cowpea > green-gram > sorghum.

An artificial diet containing teak leaf powder, Kabuligram flour and other commonly available ingredients was developed by Mathew *et al.* (1990) for rearing the teak defoliator *Hyblaea puera* (Cramer) (Lepidoptera). They reported that performance of survival in terms of per cent emergence of moth was highest on diet D1 containing Agar 20 gm, Kabuligram flour (*Cicer arietinum*) 100 gm, Casein (purified) 30 gm, Yeast extract 10 gm, Teak leaf powder 20 gm, multivitamin and mineral mixture 2 caps, Vitamins 400 mg, Ascorbic acid 3.5 gm, Sorbic acid 1 gm, Methyl parahydroxybenzote 1.5 gm, streptomycin sulphate 0.25 gm, formaldehyde 10 per cent and 2ml, distilled water 100 ml.

Singh and Ram (1987) studied the host plant spectrum in hemipterous insect, red cotton bug *Dysdercus koenigii* (Fab.) wherein the bug was reared on five different host plants *viz.,* cotton (*Gossypium hirsutum*), lady's finger (*Abelmoschus esculentus*), deccan hemp (*Hibiscus cannabinis*),

Kanghi (*Abutilon indica*) and gurhal (*Hibiscus rosasinensis*). It was observed that the different food plants had no significant effect on the development of first instar nymphs which moulted in 21 to 26 days. Deccan hemp, however, proved to be the most suitable food plant for the development of second to fifth instar hymphs, which was followed by cotton, lady's finger and Kanghi in descending order, 60, 58, 41 and 14 per cent of the nymphs became adults when they were reared on deccan hemp, cotton, lady's finger and Kanghi, respectively while, all the nymphs feeding on gurhal died before reaching the fourth instar stage.

Khan and Hajela (1987) studied the food preference and extent of damage in Coleopterous insect *Aulacophora foveicollis* (Lucas.). Their observations revealed that the first preference to *Cucurbita maxima* followed by *Cucumis sativus, Citrullus vulgaris* Var. *Lufa cylindrica* and *Lagenaria vulgaris* in descending order of preference. Multiple host test study of beetle showed that it gives first preference to *C. maxima* and last to *C. vulgaris* and *L. vulgaris*.

The experiments were carried out by Pariparna and Srivastava (1989) in dipterous fly, *Dacus cucurbitae* Coquillett to find out the optimum dose of sucrose and glucose to improve the diet. Diet having sucrose 2000 mg and glucose 500 mg per 50 ml of diet proved to be the best among different quantities tested for the growth and development of *D. cucurbitae* maggots.

Longevity and preoviposition period of *S. obliqua, A. lactinea* and *T. postica* have been studied by providing water alone, 10 per cent honey, 10 per cent sucrose, 10 per cent glucose D and protinex 5 per cent + sugar 5 per cent. Our results indicated that longevity was greatest with 10 per cent honey in all the species studied. The preoviposition period was found reduced by feeding the diet in the above species. Similar observations were noted by Sharma and Chaudhary (1985) with *H. armigera* by providing 10 per cent sugar solution.

EFFECT OF TEMPERATURE ON DEVELOPMENT AND SURVIVAL

Introduction

Under natural control, abiotic factors are perhaps the most important. As a general statement, it may be said that the insect life of a region is dependent, directly or indirectly, on the temperature, humidity, photoperiod, etc. In the study of ecology and biology of insects, temperature plays important role. The extremes of temperature limits the activities of animals and incidently determine their abundance during the annual cycle (Chapman, 1925). However, it is difficult to find out specific information dealing with its influence on species or species groups. In recent years, temperature, a climatic factor has been mostly investigated with respect to its influence on not only on the development, longevity mortality, etc., but also on the fecundity, sex-ratio, encapsulation, parasitism rate of parasitoids etc. (Mellini *et al.*, 1979). As like the temperature, humidity and photoperiod also have its own importance in population dynamics of insect world. In past, Luckman (1963), Grewal and Atwal (1969), Philipp and Watson (1971), Eubank *et al.* (1973), Bains and Shukia (1976), Thurston and Postley (1978), Kadu *et al.* (1987) and Yadava and Lal (1988) have studied the lepidopterous insects with respect to abiotic factors specially temperature and humidity. Bank and Macaulay (1970), Singh and Butter (1985), Dhiman (1986), Singh *et al.* (1986), Chandra and Kushwaha (1987), Sharma and Chaudhary (1988) and Sinha *et al.* (1990) worked ecobiologies of Hemipterous insects while, Sathe and Nikam (1981), Sathe (1985a) worked ecobiologies of Hymenopterous parasitoids; Patel and Jotwani (1986) worked on Dipterous insects with respect to ecobiology and Jagadish and Channabasavanna (1986) on the Acarines.

Materials and Methods

The laboratory reared culture of pest species were used for the experiments. Effect of temperature on development of pests from egg to adult has been studied using Refrigerators, Thermostats and B.O.D. Incubator. Experiments were conducted at constant temperatures, 10±1°C, 15±1°C, 20±1°C, 25±1°C, 30±1°C and 35±1°C. Ten individuals were subjected during each observations and it was replicated for five times. The observations were made at intervals of 12 hr. From egg hatching to formation of pupa will give larval period and from formation of pupa to adult emergence pupal period.

Results

S. obliqua

There was no egg hatching at 10°C and 35°C (Table 18). At 30°C egg hatching period was minimum and found progressively prolonged with the temperature 25°C, 20°C and 15°C. Larval development was also fast at 30°C at which it took 21 days. At 25°C, 20°C and 15°C, the average period of larval life was 37, 42 and 51 days respectively (Table 19). At 10°C and 35°C larvae could not proceed with their developments. Percentage of survival was also found optimum at 30°C. Pupal development averaged 21.4, 13.2, 11.4 and 7.2 at 15, 20, 25 and 30°C respectively. In general, 30°C was the optimum temperature for the development of *S. obliqua* (Table 20, Figures 56 and 57).

A. lactinea

Results are shown in Tables 18, 21 and 22. At 15, 20, 25 and 30°C egg hatching periods, larval periods and pupal periods were 7, 5, 4, 3; 48.4, 34.2, 27.8, 23 and 20.4, 16.2, 12.4 and 8 days respectively. There was no egg hatching and larval and pupal development at 10° and 35°C (Figures 58–59).

T. postica

As like the above pest species, development of eggs, larvae and pupae could not seen at the temperatures 10°C and 35°C (Tables 23 and 24). The life cycle of *T. postica* was completed within 57.8. 43.1. 32.1 and 25.2 days at 15, 20, 25 and 30°C respectively. The maximum percentage of survival and the formation of individuals was also found at 30°C (Figures 60 and 61). On the basis of above data the most favourable temperature for development of all the species seemed to be 30°C.

Discussion

Investigations were made to determine the effect of abiotic factor, temperature on development and survival of *S. oblioua, A. lactinea* and *T. postica.* The main object of the present study was to find out the optimum temperature for fast development of the pest species. The optimum conditions are essential for mass multiplication of the pest and in a pest control programme mass rearing of host has immense value.

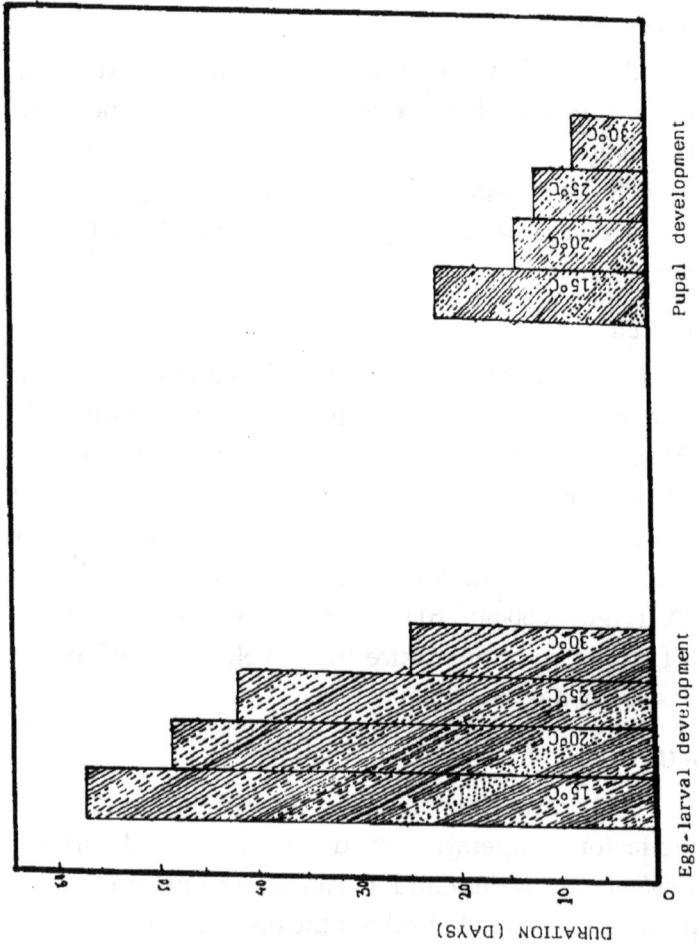

Figure 56: Effect of Temperature on Egg-Larval and Pupal Development of *S. obliqua*

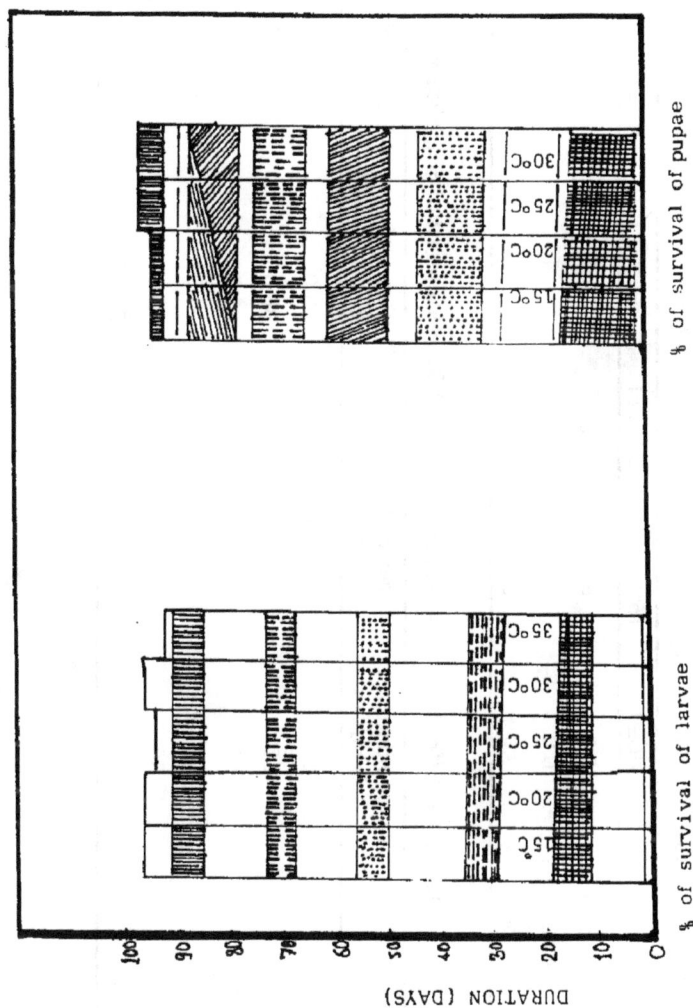

Figure 57: Effect of Temperature on Percentage of Survival of Larvae and Pupae of *S. obliqua*

Table 18: Egg Hatching in Relation to Different Constant Temperature

Sl.No.	Pest Species	Mean Number of Days Required for Hatching					
		10°C±1	15°C±1	20°C±1	25°C±1	30°C±1	35°C±1
1.	S. obliqua	0.0	6	5	4	3	0.0
2.	A. lactinea	0.0	7	8	4	4	0.0
3.	T. postica	0.0	6	4.5	3.5	3	0.0

Table 19: Effect of Temperature on Larval Development of S. obliqua

Temp. ±1°C	No. of Individuals Tried	No. of Adults Formed	Per cent of Survival	Average Period (Days)
10	50	46	0.0	Nil
15	50	48	96.00	51.00
20	50	48	96.00	43.00
25	50	47	94.00	37.5
30	50	48	96.00	21.5
35	50	46	92.00	0.0

Table 20: Effect of Temperature on Pupal Development of *S. obliqua*

Temp. ±1°C	No. of Pupae Tried	No. of Adults Emerged	Per cent of Survival	Average Period (Days)
10	50	NIL	NIL	NIL
15	50	47	94.00	21.5
20	50	47	94.00	13.5
25	50	48	96.00	11.5
30	50	48	96.00	7.5
35	50	NIL	NIL	0.0

Table 21: Effect of Temperature on Larval Development of *A. lactinea*

Temp. ±1°C	No. of Larvae Tried	No. of Pupae Formed	Per cent of Survival	Average Period (Days)
10	50	NIL	NIL	NIL
15	50	46	92.00	42.5
20	50	45	90.00	35.5
25	50	47	94.00	31.5
30	50	49	98.00	27.0
35	50	NIL	Nil	NIL

Table 22: Effect of Temperature on Pupal Development of *A. lactinea*

Temp. ±1°C	No. of Pupae Tried	No. of Adults Formed	Per cent of Survival	Average Period (Days)
10	50	NIL	Nil	NIL
15	50	48	96.00	24.5
20	50	47	94.00	20.5
25	50	.49	98.00	16.5
30	50	49	98-00	12.0
35	50	NIL	NIL	NIL

Table 23: Effect of Temperature on Larval Development of *T. postica*

Temp. ±1°C	No. of Larval Tried	No. of Pupal Formed	Per cent of Survival	Average Period (Days)
10	50	Nil	Nil	NI
15	50	47	94.00	33.5
20	50	46	92.00	25.0
25	50	48	96.00	18.5
30	50	49	98.00	15.2
35	50	Nil	Nil	Nil

Table 24: Effect of Temperature on Pupal Development of *T. postica*

Temp. ±1°C	No. of Pupae Tried	No. of Adults Emerged	Per cent of Survival	Average Period (Days)
10	50	Nil	Nil	Nil
15	50	48	96.00	18.5
20	50	48	96.00	14.00
25	50	48	96.00	10.5
30	50	50	100.00	7.0
35	50	Nil	Nil	Nil

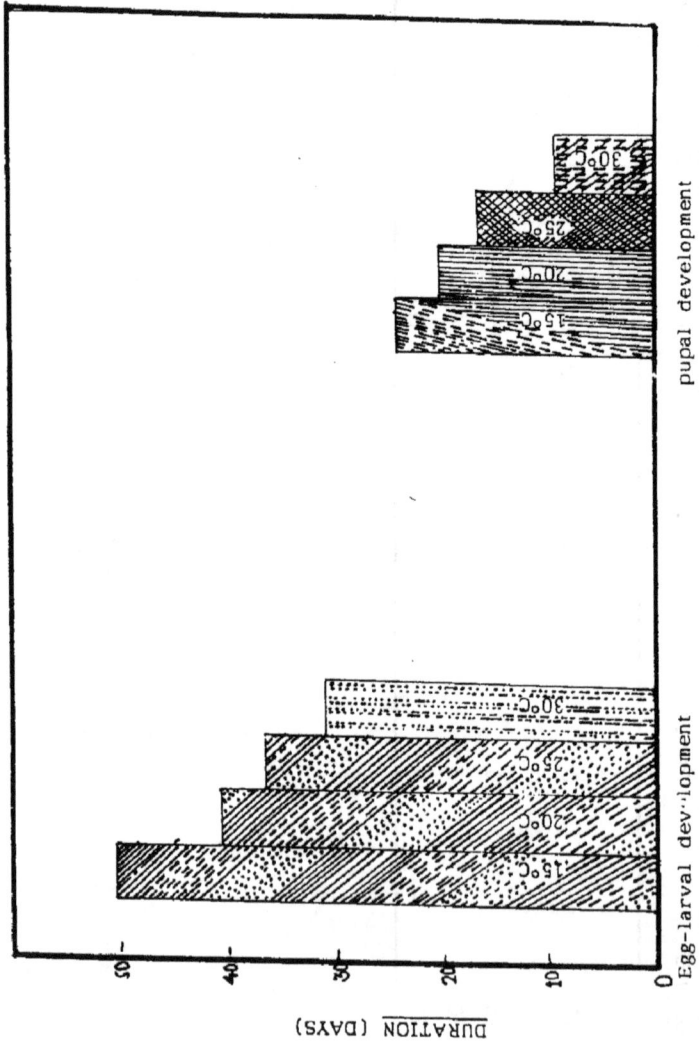

Figure 58: Effect of Temperature on Egg-Larval and Pupal Development of *A. lactinea*

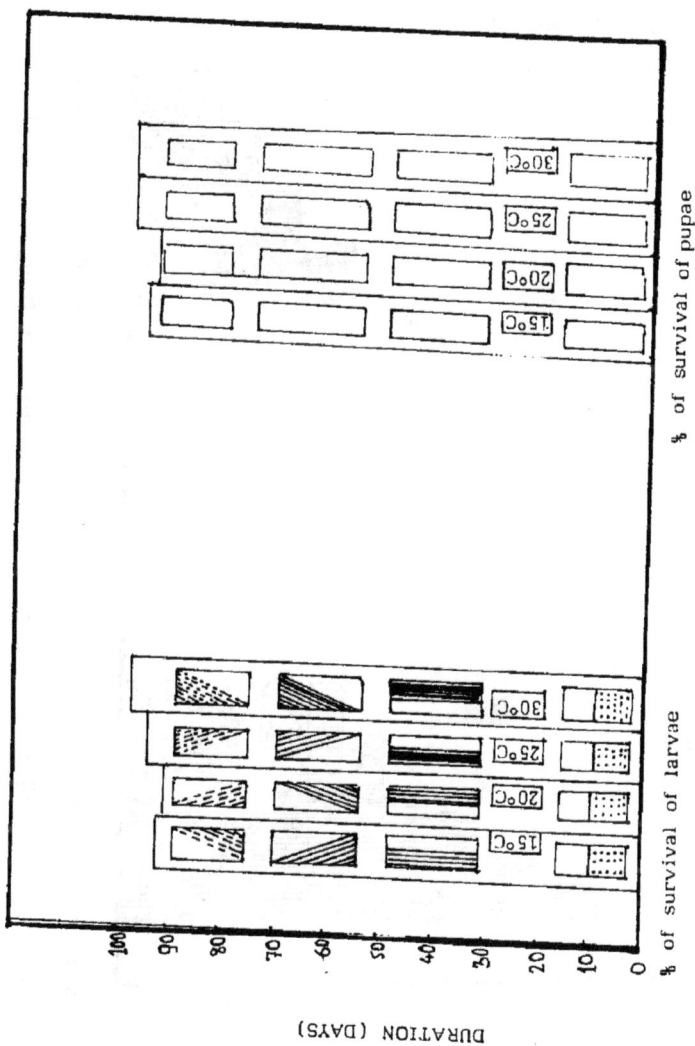

Figure 59: Effect of Temperature on Percentage of Survival of Larvae and Pupae of *A. lactinea*

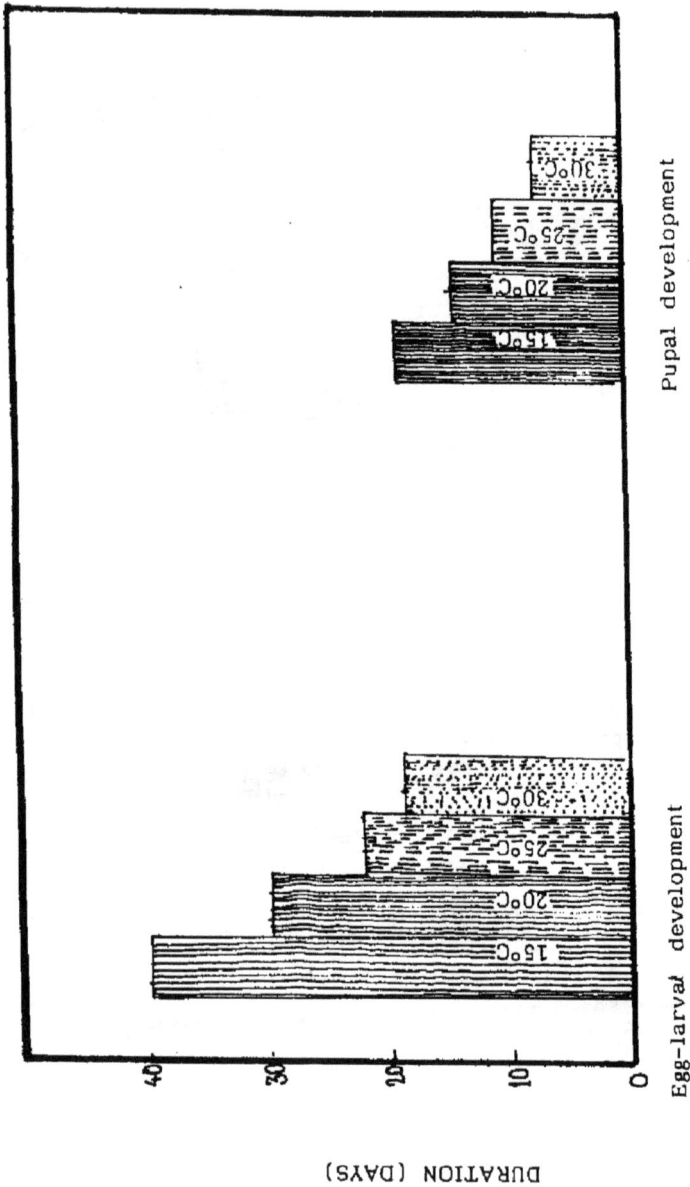

Figure 60: Effect of Temperature on Egg-Larval and Pupal Development of *T. postica*

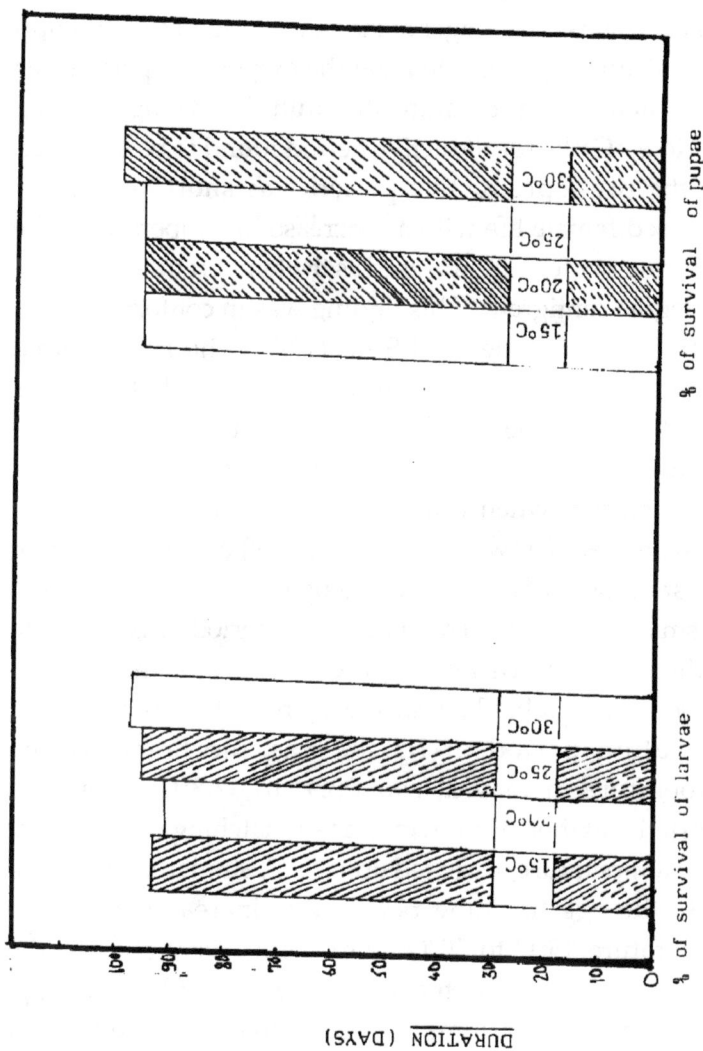

Figure 61: Effect of Temperature on Percentage of Survival of Larvae and Pupae of *T. postica*

The Lepidopterous pests of agricultural importance have been studied by several workers with respect to abiotic factor and population dynamics. Grewal and Atwal (1969) studied the cotton boll worm *Earias insulana* Stoll. They observed that 8 days male life of *E. insulana* at 25°C, and female life was little longer at the same temperature. Philipp and Watson (1971) reported that the oviposition period was longest at lower temperature, the adults lived longer during 20.6° to 23°C, the reproductive period found to be increased and the pre-reproductive period was short because of decreased female life with an increase in temperature. The egg laying per female increased as the temperature increased per degree centigrade. This finding was in conformity with the findings of Butter and Scott (1976) who has reported that egg laying of *H. zea* were 10 worms per 100 earheads to 1–2 worms per earhead, where the temperature increased from Ca 900 heat units to Ca 1300 heat units indicating that when heat units increased, the egg laying also increased. Grewal and Atwal (1969) also reported that largest number of eggs were laid by *E. insulana* at 30°C and the smallest at 25°C. The incubation period was found to be decreased with per degree centigrade increase in temperature. Eubank *et al.* (1973) reported that the egg incubation period of *H. zea* was shortened as the temperature increased from 23.9° to 32.3°C. They also reported that there was an increase in hatching of *E. insulana* as the temperature increased upto 30°C. Similarly, in present study the egg hatching period was increased from the temperature 15°C to 30°C. Luckmann (1963) studied the rate of development of the incubating eggs of *Heliothis* spp. at different temperatures. He found that the development began at 54°F. The negative relationship was observed between 1st and 4th instar larval developmental period and temperature (Kadu *et al.*, 1987). Komrova (1959) observed that the normal pupal period 10 to 20 days of *H. armigera* at 23°–25°C and the pupal period extended if the pupae kept at low temperature. Grewal and Atwal (1969) also reported that the pupal stage of *E. insulana* decreased with per degree centigrade increase in temperature. Philipp and Watson (1971) recorded that the mean generation time was shortest for pink bollworm at higher temperature such as 83°F.

Fye and Poole (1971) studied the effect of high temperatures on the fecundity and fertility of six lepidopterous pests of cotton in Arizona wherein they noted decreasing longevity and fecundity at the constant temperature regimes as the temperature increased.

Fye and May (1974) studied the development, fecundity and longevity of the cotton leaf perforator *Bucculatrix thurberiella* Busck wherein the developmental periods for egg larval and pupal stages were 9.2±0.5, 25.5±1.1, 12.8±0.7; 6.3±0.6, 14.4±1.1, 6.1±0.8; 4.5±0.4, 10.3±0.6, 4.3±0.4 and 3.8±0.1, 10±0.6, 3.9±0.2 at 20, 25, 30 and 35°C respectively.

Kadu *et al.* (1987) studied the effect of temperature on the development of *Helicoverpa armigera* (Hubner) (Lepidoptera). They observed that the adult female life, male life, pre-reproductive period and incubation period was shortened with the increase in temperature. The reproductive period, number of eggs laid per female and hatching percentage of eggs increased with the rise in temperature per degree centigrade. The reproductive period of female increased when the average temperature ranged from 23.0°C to 27.8°C. The hatching percentage of eggs was more when temperature ranged between 25.4°C to 27.8°C. It took less period to complete the development of 1st and 4th instar larvae with per degree centigrade increase in temperature while, developmental period of 2nd and 3rd instar larvae and total larval developmental period increased with per degree centigrade rise in temperature. The pupal and generation period shortened with the increase in temperature.

Again Sharma and Chaudhary (1988) studied the above insect with respect to effect of 4 levels of constant temperature and 3 levels of constant relative humidity on rate of development of different immature stages of larvae. The average incubation periods at 20, 25, 30 and 35°C was 5, 4, 3, and 2 days respectively. The larval periods at the above four constant temperatures were 31.4, 19.3, 15.3 and 10.3 days, pupal periods 23.8, 15.5, 9.6 and 8.4 days and the total developmental durations 62.2, 38.8, 27.8 and 20.7 days respectively and there was a negative correlation between these four constant temperatures and duration of different immature stages. The hatchability of eggs at 20, 25, 30 and 35°C were 63.1, 69.8, 74.7 and 84.2 per cent respectively, showing a positive correlation between temperature and hatchability of eggs. The different level of constant humidity from 30 to 80 per cent did not gave any adverse effect on the rate of development as well as on the viability of eggs. The pupal duration, however, was slightly enhanced with the increase of humidity at 20 and 25°C. It was 23 days at 30 per cent RH- and 25.2 days at 80 percent RH- at 20°C and 13.5 days at 30 per cent RH- and 17 days at 80 per cent RH- at 25°C. However, at 30 and 35°C there was no appreciable difference in the pupal durations at the above three levels of humidity. In the present study, egg hatching periods, larval periods and pupal periods were 6, 5, 4 and 3 days; 51.43, 37.5 and 21.5 days and 21.5, 13.5, 11.5 and 7.5 days at 15, 20, 25 and 30°C respectively in *S. obliqua*. The total developmental time from egg to adult emergence was averaged 78.5, 61.5, 53 and 32; 76, 56, 44.5 and 34 and 58, 43.5, 32.5 and 25.5 days at 15, 20, 25 and 30°C in *S. obliqua, A. lactinea* and *T. postica* respectively.

Singh *et al.* (1986) studied the abiotic factors with respect to population of mustard aphid, *Lipaphis erysimi* (Kalt) and green peach aphid, *Myzus persicae* (Sulzer) (Hemiptera). They found that maximum and minimum temperature and sunshine had significantly favourable influence, morning and evening relative humidity had direct positive effect on a low magnitude and wind velocity and rainfall were found to have direct negative effect on the population of *L. erysimi*. Morning and evening relative humidity had significantly negative influence, minimum temperature and rainfall had direct negative effect and maximum temperature, wind velocity and sunshine had direct positive effect on the population of *M. persicae*. The present study was carried out at laboratory conditions at different constant temperatures and the field abiotic parameters were not considered.

Dhiman (1986) studied the effect of temperature on the seasonal occurrence of *Cletus signatus* Walker (Hemiptera) under field condition. He noted that the suitable temperature for the breeding of this bug was 32.4 to 34.1°C and the maximum number of the bug has been observed during June to September. In winter months, December to middle of February, it hibernated in adult stage as the temperature falls considerably.

On *Brassica* spp. under field condition Sinha *et al*. (1990) studied the population dynamics of mustard aphid *L. erysimi* in relation to ecological parameters. He reported that the environmental parameter play an important role in aphid infestation. The ambient maximum (21.68 to 23.52°C), and minimum (7.18 to 9.40°C) temperatures in January, February, appeared to be most conducive for the aphid multiplication. High humidity had little impact on aphid population fluctuation. Minimum humidity, ranging from 55.7 to 69.4 per cent in January–February favoured the population build up while, the activity of the aphid ceased at 50.90 per cent and below.

Patel and Jotwani (1986), reported that the average mean temperature between 25 to 30°C and RH above 60 per cent favoured increased midge activity. Jagadish and Channabasavanna (1986) studied the development of *Typhlodromips tetranychivours* Gupta (Acari.) with respect of temperature and humidity and found that the incubation period of eggs was minimum (15.87±1.22 hr) at 24±1°C and maximum (79.85±28.54 hr at 20±1°C with humidity combination of 75±3 per cent and 95±3 per cent respectively. Optimum range of temperature and humidity for the egg hatchability was 27±1°C to 30±1°C and 65±3 per cent and 85±3 per cent respectively. Optimum temperature and humidity for development was noted to be in the range 24 to 27±1°C and RH- 65 to 95±3 per cent.

Nikam and Sathe (1981) studied the development and survival of an Ichneumonid *Diadegma trichoptilus* Cameron (Hymenoptera), a larval parasitoid of *Exelastis atomosa* Walsinham. They noted that the duration for complete development was 35, 31, 27, 19, 15 and 13 days at 10, 15,

20, 25, 30 and 35°C respectively and the percentage of survival was 33.33, 53.33, 76.63, 89.99, 63.33 and 30.00 at the above temperatures respectively. The optimum temperature for development was 25°C. In *Cotesia diurnii* Rao and Nikam, a larval parasitoid of *E. atomosa* there was no larval as well as pupal development at 10°C and 35°C. However, the development was completed in 25.10, 23.39, 15.30, 12.50 and 11.90 days at 15, 20, 25, 30 and 32°C respectively. The optimum temperature for development was 25°C with high percentage of survival. While, in the present study the optimum temperature for development was 30°C, the percentage of survival was also higher at this temperature in all three species studied.

LONGEVITY, FECUNDITY AND SEX-RATIO

Introduction

Evaluation of biological control is one of the most difficult task, since most biological control projects are unique experiments in community relationship and difficult to repeat. Biological control is widely accepted as an alternative approach to pesticidal measures. The workers are increasingly attracted in doing research on bio-control of several crop pests. Hence, biological data on bio-control agents and pest species is must for the assignment of bio-control technique. Specially basic population processes of the species like ovipositional rates, developmental rates, longevity, etc. should be studied thoroughly. Since fecundity and sex ratio are the important components of mass rearing of the species, the present study was carried out on three pests *viz.. S. obliqua, A. lactinea* and *T. postica*. The fecundity data is useful for constructing life-tables of the insects and also to calculate the intrinsic rate of increase in the species.

Materials and Methods

The culture of *S. obliqua, A. lactinea* and *T. postica* were maintained in laboratory by collecting the larvae of the above species from the fields of sunflower, soyabean, cowpea etc. at Kolhapur. Laboratory reared insects were used for the experiments. Newly emerged male and female of the species were caged in a glass cage (Figures 3, 4) along with the green leaves of the host plants. Eggs were laid by the females on the leaves or sometimes on the paper in the cage. Eggs were collected with the help of hair brush and kept in the containers for further development upto the adult. A single mated female was allowed to lay her eggs till the death for studying the fecundity. With the help of progeny production, fecundity was observed. The experiments were replicated for 10 times.

Results

S. obliqua

The longevity of female varied from 4–5 (average 4.8) days. Total number of eggs deposited varied from 352 to 457 (average, 400.2) and the total number of individuals produced ranged from 347 to 450 (average, 392.9) with an average sex-ratio, male : female, 1 : 0.991 (range, 1 : 0.487 to 1 : 1.657). The results are recorded in Table 25 and Figure 62.

A. lactinea

The longevity of female moths was varied from 4–6 days (average 5.1 days). The maximum number of eggs laid was 288 and the minimum, 199 (average 241.00). The total number of progeny production averaged 231.2 individuals (range, 280–184 adults). The sex ratio, male : females ranged

Table 25: Fecundity, Longevity and Sex-ratio of *S. obliqua*

Female No.	Total No. of Eggs Laid	Longevity of Females	Total Number of Individual Produced			Sex-ratio
			Male	Female	Total	Male : Female
A	352	4	148	199	347	1 : 1.344
B	411	5	174	281	405	1 : 1.327
C	414	5	275	134	409	1 : 0.487
D	388	5	187	194	381	1 : 1.037
E	381	5	171	206	377	1 : 1.204
F	457	5	270	180	450	1 : 0.666
G	417	5	152	252	404	1 : 1.657
H	387	5	191	183	374	1 : 0.958
I	418	5	211	199	410	1 : 0.943
J	377	4	194	178	372	1 : 0.896
Average	400.2	4.8	197.3	195.6	392.9	1 : .991

Table 26: Fecundity, Longevity and Sex-ratio of A. *lactinea*

| Female No. | Total No. of Eggs Laid | Longevity of Females | Total Number of Individual Produced | | | Sex-ratio |
			Male	Female	Total	Male : Female
A	288	6	200	71	271	1 : 0.355
B	211	5	150	51	201	1 : 0.34
C	214	5	97	102	199	1 : 1.05
D	282	6	126	154	280	1 : 1.22
E	262	5	129	128	257	1 : 0.99
F	259	5	104	137	241	1 : 1.317
G	200	5	88	106	194	1 : 1.20
H	199	4	99	85	184	1 : 0.85
I	271	6	205	61	266	1 : 0.297
J	224	4	149	70	219	1 : 0.46
Average	241.0	5.1	134.7	96.5	231.2	1 : 0.716

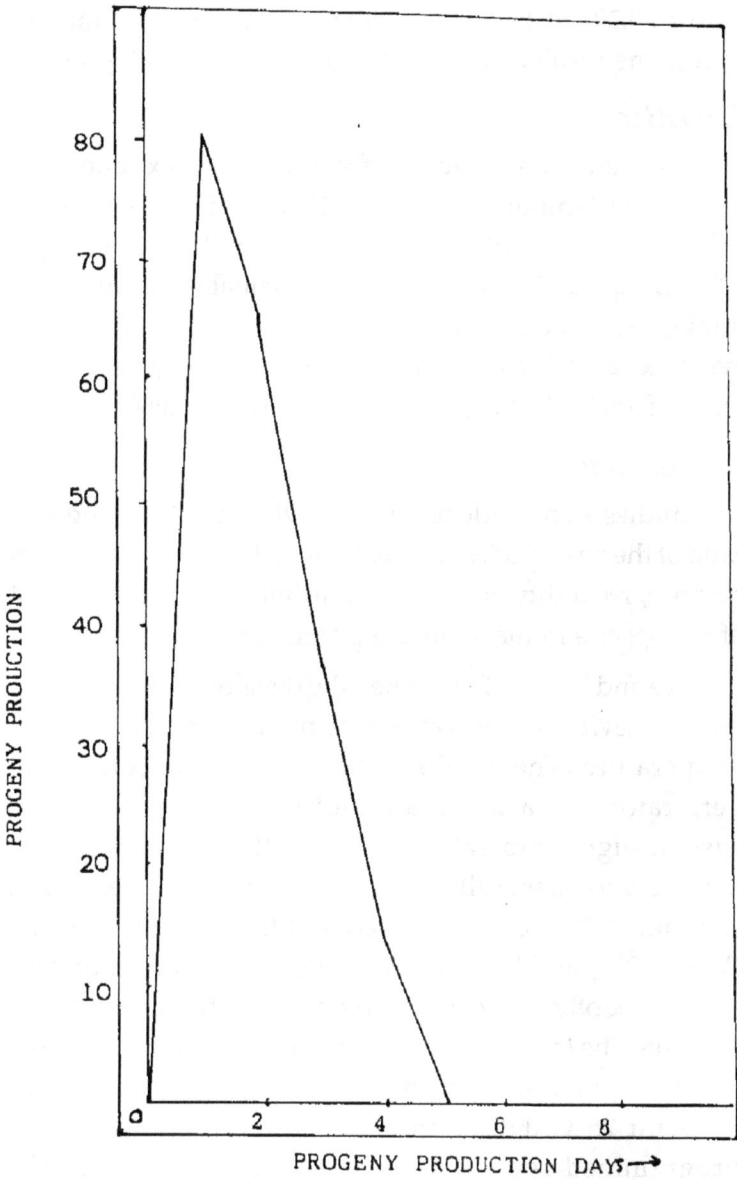

Figure 62: Relationship between Progeny Production and Progeny Production Days of *S. obliqua*

from 1 : 0.297 to 1 : 1.317 (average. 1 : 0.716) favouring the males. The results are tabulated in Table 26 and Figure 63.

T. postica

The results on longevity, fecundity and sex ratio of *T. postica* are tabulated in Table 27 and Figure 64 which indicated that the total number of eggs laid were averaged 305.5 (range 287–319), longevity of females averaged 4.6 (range, 4–5) and the total number of individuals produced averaged 299.9 (range, 281–314) with an average sex-ratio, male : female, 1 : 0.962 (range 1 : 0.775–1 : 1.666).

Discussion

Studies were made on the longevity, fecundity and sex-ratio of the pest species *i.e. S. obliqua, A. lactinea* and *T. postica* because, fecundity is desirable attribute for the assessment of the species in mass culture programme.

Fye and May (1974) studied the development, fecundity and longevity of the cotton leaf-perforator in relation to temperature. The results indicated that the cotton leaf perforator was a warm and not hot weather insect. In sustain high temperature regimes, the fecundity of the females was drastically reduced. The females reared at a constant, 35°C were short lived and laid no eggs. Fye and Poole (1971) and Fye and McAda (1972) observed parallel data in six other species of lepidoptera attacking cotton in Arizona. The longevity of the male moths and the fecundity in the programmed 21°C regime closely approximated the constant 25°C data; although the females in the 21°C programmed regime lived nearly twice as long. The longevity of the individuals in the greenhouse study fell midway between the individuals in the 20°C and 25°C

Table 27: Fecundity, Longevity and Sex-ratio of *T. postica*

Female No.	Total No. of Eggs Laid	Longevity of Females	Total Number of Individual Produced			Sex-ratio
			Male	Female	Total	Male : Female
A	287	4	147	134	281	1 : 0.911
B	319	5	161	153	314	1 : 0.916
C	307	5	156	143	299	1 : 0.950
D	314	5	174	135	309	1 : 0.775
E	298	4	144	149	293	1 : 1.347
F	294	5	131	162	293	1 : 1.236
G	312	5	138	161	299	1 : 1.666
H	302	4	149	145	294	1 : 0.973
I	309	4	164	142	306	1 : 0.865
J	317	5	164	147	311	1 : 0.896
Average	305.5	4.6	152.8	147.1	299.9	1 : 0.962

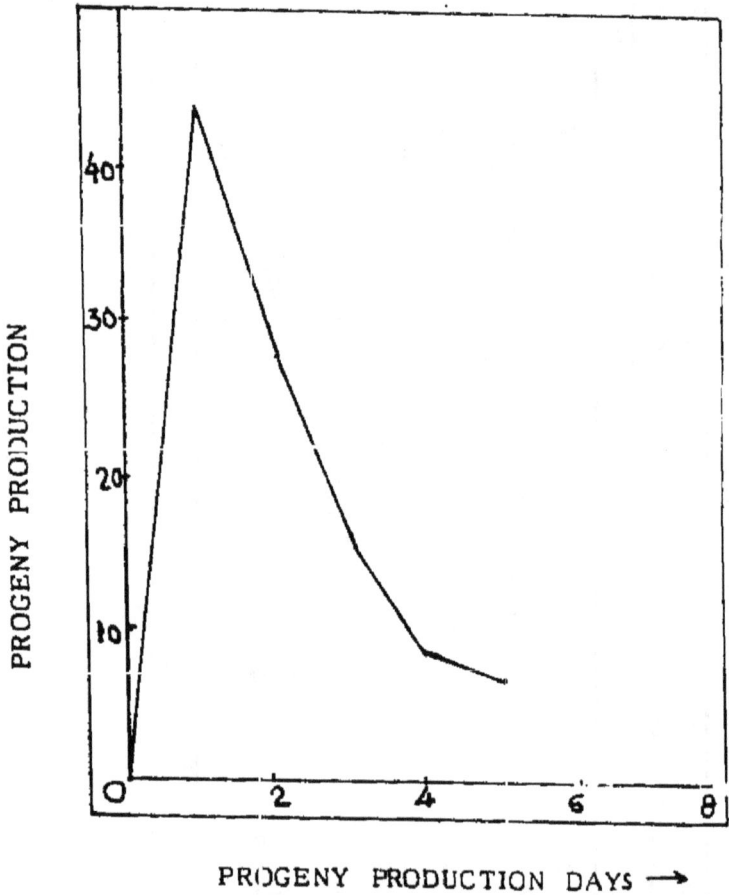

Figure 63: Relationship between Progeny Production and Progeny Production Days of *A. lactinea*

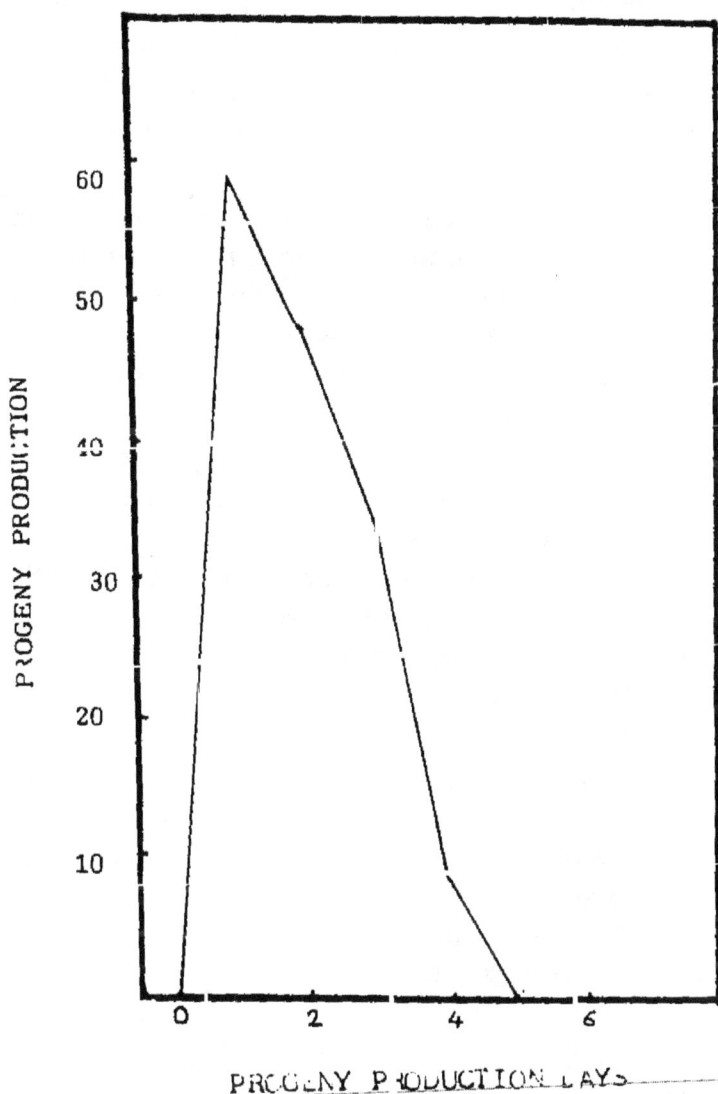

Figure 64: Relationship between Progeny Production and Progeny Production Days of *T. postica*

constant regimes. However, the fecundity more closely approached the fecundity of the females in the constant 20°C. Although, the daily maximum temperatures included in the greenhouse regime were relatively high, the higher relative humidity in the greenhouse apparently had a marked favourable effect on the fecundity.

Ellington (1970) made evaluation studies with respect to host plant resistance to lepidoptera. He has suggested the same phenomenon for the boll-worm *Heliothis zea* (Boddie) and for the cabbage lopper *Trichoplusiani* (Hubner) as suggested by Fye and Poole (1971). Ellington (1970) found that the fecundity was drastically reduced by low humidities and that in high humidities the potential of the ovipositioning females was better achieved. The relative low fecundity in 21°C programme regimes may have been partially due to the constant disturbance of the ovipositioning females. The fecundity data of Fye and May (1974) agree with that of Watson and Johnson (1972) only at 25°C. However, at 35°C the fecundity was reduced to zero. In present study the fecundity was studied at room temperature (25±1°C, 55–60 per cent RH, 12 hr photoperiod) and the results showed that in *S. obliqua* females on an average produced 392.9 adults with sex-ratio, male : female, 1 : 0.991. The sex ratio of the progeny production was vary. In *A. lactinea* the progeny production ranged from 184–280 adults (average, 207.1) with (male : female) sex-ratio, 1 : 0.666. The results on the fecundity of *T. postica* indicated that the progeny production averaged 299.9 individuals with sex ratio, male : female, 1 : 0.962.

Fecundity in parasitic hymenoptera have been widely attempted by several authors (Broodryk, 1969a;

Oatman and Platener, 1974; Sato, 1975; Yeargan *et al.*, 1978 and Sathe, 1987). Broodryk (1969b) studied the fecundity in a Braconid wasp *Orgilus parcus* Turner in which he found that the progeny production was 64.8 individuals and the sex-ratio, male : female 2.2 : 1. Oatman and Planter (1974) observed the fecundity of *Temelucha* sp. (*Platensis* group) in which they noted that the fecundity was very high compare to the fecundity of *O. parcus*. Cardona and Oatman (1971) studied the fecundity in mated and virgin females and reported mean progeny 152.1 which was higher than that for virgin females, 120.1 in *Pseudapanteles dignus* Muesebeck (=*Apanteles dignus*). In an another Braconid, *Cotesia glomeratus* Say Sato (1975) found that the percentages of emergence of parasitoid larvae, *C. glomeratus* from the host *Pieris rapae cracivora* feeding on cabbage leaves and the artificial diet were 93 and 89 respectively. Yeargan and Latheef (1977) and Yeargan *et al.* (1978) studied the fecundity, oviposition and longevity of *Bathyplectes anurus* Thompson and *B. curculionis* (Thompson) in which they reported that ovipositional rate and longevity were influenced by temperature fluctuations but fecundity was not affected significantly. Whereas, the present study was conducted under laboratory conditions and room temperature with lepidopterous insects *i.e. S. oblioua*, *A. lactinea* and *T. postica* and their fecundities were higher.

The data will serve as a basis for mass rearing of the above pest species and also for constructing life tables of the species and further calculating the intrinsic rates of natural increase.

LIFE TABLES AND INTRINSIC
RATES OF INCREASE

Introduction

In the population ecology we must deal with organisms capacity for increase in numbers, 'r_m' which is the inherent capacity for increase of population of stable age distribution under given physical conditions when competition and other biotic mortality factors are absent. The actual rate of increase, 'r', is the rate of increase that pertains under any prevailing condition, and is a more real statistic for population assessment. The intrinsic rate is useful in comparing the potentials of different pest species, or of a pest species and the natural enemies of pest. Life table studies giving fecundity and survivalship data, combined with statistical determinations enable us to evaluate these various population growth or decline parameter in relation to various environmental factors that affect the population (Deevey, 1947; Birch, 1948). The equilibrium mortality is exceeding important in pest population increase or decline. It merits more attention.

S. obliqua, A. lactinea and *T. postica* are important agricultural and horticultural pests in India. The control of the above pests is a pressing problem. Hence, the essential objects are to make the estimates of the rate of growth of these pests and their natural enemies. As far as life tables and intrinsic rate is concerned Thompson (1924) for the first time developed a mathematical method. However, this method proved to be very much laborious. Later, Lotka (1925) derived a function for the intrinsic rate of natural increase 'r_m'. Lotka designed this function for human population and for the first time Birch (1948) extended it into insect population. Review of literature indicates that life-table studies have been attempted in different orders of insects by several ecologist and entomologist amongest whom the notable works may be mentioned those of Morris and Miller (1954), Stark (1959), Waloff (1968), LeRaux *et al.* (1963), McLeod (1972), Bains and Shukla (1976), Bilapate and Pawar (1980), Nikam and Sathe (1983a) and Sathe (1986a, 1988a). In the present study, the life tables were constructed according to Birch, (1948) as elaborated by Howe (1952) and Watson (1964).

Explanation of Birch's Method

Birch (1948) says that the intrinsic rate of natural increase is the actual rate of increase of population under specified constant environmental condition where space and food are unlimited and no mortality factors are present other than physiological:

$$\Sigma\ e^{-r}m^x lxm_x = 1$$

where,

- e: Is the base of the natural longarithms
- x: The age of the individual in days.
- lx: The number of individuals alive at age x as a proportion of one.
- m: The number of female offsprings produced per female in the age interval 'x'.

The sum of products $l_x m_x$ is the net reproductive rate, 'Ro', which is the rate of multiplication of the population in each generation measured in terms of females produced per generation.

The approximate value of cohort generation time 'T_c' was calculated as follows:

$$T_c = \frac{l_x m_x X}{l_x m_x}$$

For arbitary value of innate capacity for increase 'r_c' was calculated from the following formula:

$$r_c = \frac{\log_e R_0}{T_c}$$

The value of r_m upto two decimal places was added in the formula until two values of the equation were found which lies immediately above or below 1096.6.

On the vertical axis abritary 'r_m's were taken while on the horizontal axis two values of

$$\Sigma e^{7-r} m_x l_x m_x = 1$$

were plotted. The two points were joined to give a line which intersected a vertical line drawn from the desired value of 'r_m' accurate to 3 decimal places. The precise generation time 'T' was then calculated from the formula:

$$T = \frac{\log_e R_0}{r_m}$$

The finite rate of increase (λ) was calculated as

$$\lambda = e^r m$$

Materials and Methods

Rearing of the pest species namely *S. obliqua, A. lactinea* and *T. postica* were maintained in the laboratory as per the procedure given in material and methods. After mating of the pest species moths were kept for egg laying in glass jars/glass cages (Figures 3 and 4). Newly emerged eggs were transferred daily into the plastic containers and petridishes for larval development. The larvae were fed with their natural food material. After the formation of pupae, they were separated in a large containers for adult emergence. Thus, observations were made on their progeny production of mated females. Later, sex-ratio was calculated. After collecting fecundity data, life-tables were constructed according to Birch (1948), and intrinsic rate of increase was calculated in the species.

Results

S. obliqua

Longevity of ovipositing females ranged from 4 to 5 days (average, 4.8 days). The number of progeny production averaged 392.9 (range, 347–450) adults with a sex-ratio, male : female, ranged from 1 : 0.487 to 1 : 1.657 (average 1 : 0.991). Average length of immature forms was 31 days at 30°C. The maximum mean progeny production per day, 'm_x' was 80.6 on the first day and reproduction stopped on 5th day in a single generation. The intrinsic rate of increase in the species was 0.163 (Figure 65) per female per day and population multiplied to 174.42 times in mean generation time, 'T' of 31.66 days. The statistical results are represented in Tables 28–31.

$$T_c = \frac{l_x m_x X}{l_x m_x} = \frac{5692.78}{174.42} = 32.63$$

where T_c *is arbitrary T*

$$r_c = \frac{\log_e R_0}{T_c} = \frac{\log_e 174.42}{32.63} = 0.158$$

where r_c *is arbitrary* r_m

Now $T_c = 32.63$, $r_c = 0.158$,

Now, arbitrary 'r_m's (r_c) are 0.13 and 0.17

∴ $r_m = 0.163$ (Figure 65)

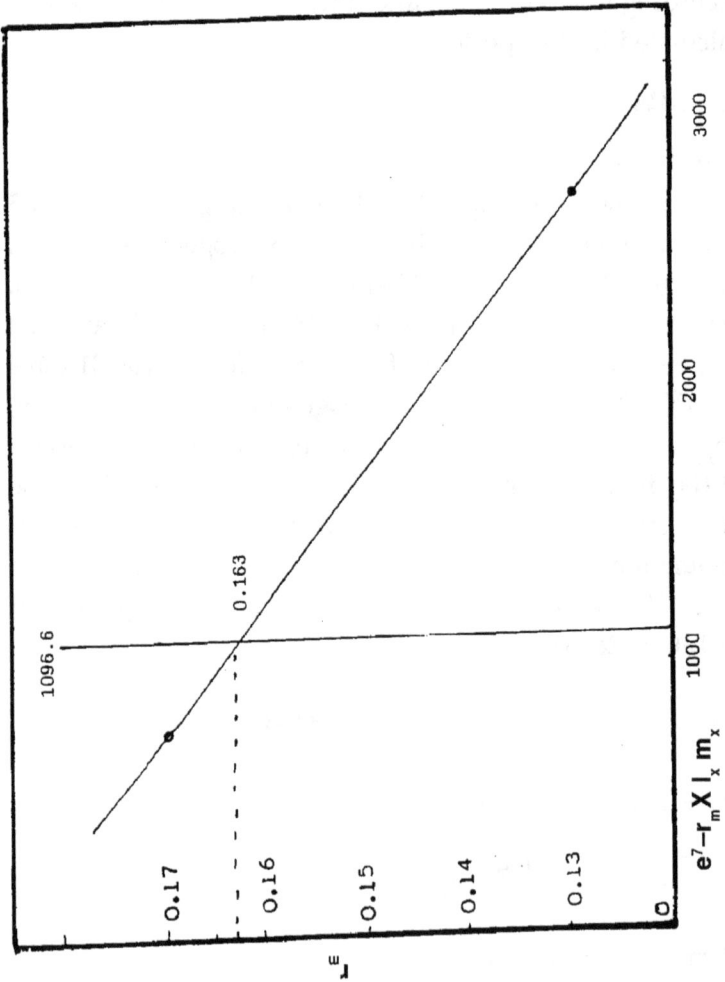

Figure 65: Determination of Intrinsic Rate of Increase in *S. obliqua*

Table 28: Daily Female Progeny Production by Mated Females of S. obliqua

Female No.	Number of Females/Day					Total Individuals Produced
	1	2	3	4	5	
A	84	75	40	D	–	199
B	90	79	52	10	D	231
C	71	60	3	D	D	134
D	71	60	51	12	D	194
E	94	71	31	10	D	206
F	64	49	40	27	D	180
G	97	92	40	23	D	252
H	94	59	30	D		183
I	62	51	45	41	D	199
J	79	47	37	15	D	178
Average	80.6	64.3	36.9	13.8	0	195.6

where λ is finite rate of natural increase

$$T = \frac{\log_e 174.42}{0.163} = 31.66 \, \text{days}$$

Table 29: Life Table Statistics of *S. obliqua*

Pivotal Age Days X	Proportional Life at Age x l_x	No. of Female Progeny/Female m_x	$l_x m_x$	$l_x m_x X$
32	1	80.6	80.6	2597.2
33	1	64.3	64.3	2121.9
34	0.8	36.9	29.52	1003.68
35	0	13.8	0.00	000.00
			174.42	5692.78

Table 30: Provisional r_m (0.13) for *S. obliqua* and Related Values of $7^{-r}m_x \, 1_x m_x$

x	$r_m X$	$7^{-r}m_x$	$7^{-r}m_x$	$7^{-r}m_x l_x m_x$
32	4.16	2.84	17.115	1379.469
33	4.29	2.71	15.029	966.364
34	4.42	2.58	13.197	389.575
35	4.55	2.45	11.588	000.000
				2735.408

Table 31: Provisional r_m (0.17) for *S. obliqua* and Related Values of $7^{-r}m_x \, 1_x m_x$

x	$r_m X$	$7^{-r}m_x$	$7^{-r}m_x$	$7^{-r}m_x l_x m_x$
32	5.44	1.56	4.758	383.494
33	5.61	1.39	4.0148	258.100
34	5.78	1.22	3.3871	99.984
35	5.95	1.05	2.8576	00.000
				741.578

A. *lactinea*

The duration for immature stages was 41 days, ovipositing female's longevity averaged 5.1 days (range, 4–6). The progeny production averaged 231.2 individuals with an average sex-ratio (male : female) 1 : 0.716. The 'm_x' *i.e.* maximum mean progeny production per day was 43.8 on first day and progeny production stopped on the 5th day. The intrinsic rate of increase was 0.107 (Figure 66) and population multiplied to 93.36 in a generation time of 42.39 days. The statistical results are represented in Tables 32–35.

$$T_c = \frac{l_x m_x X}{l_x m_x} = \frac{3999.74}{93.36} = 42.84$$

where T_c is arbitray T

$$r_c = \frac{\log_e R_0}{T_c} = \frac{\log_e 93.36}{42.84} = 0.1058$$

where r_c is arbitray r_m and arbitrary 'r_m's

(r_c) are 0.11 and 0.09

∴ r_m = 0.107 (Figure 66)

$$T = \frac{\log_e 93.36}{0.107} = 42.39 \, \text{days}$$

T. *postica*

The duration of immature stages in *T. postica* was 25 days. Longevity of ovipositing females averaged 4.6 days (range, 4–5 days). The progeny production averaged 299.9 individuals (range, 281–314 individuals). The offsprings (male : female) averaged 1 : 0.962 (range, 1 : 0.775–1 : 1.666). Maximum mean progeny production per day 'm_x' was 59

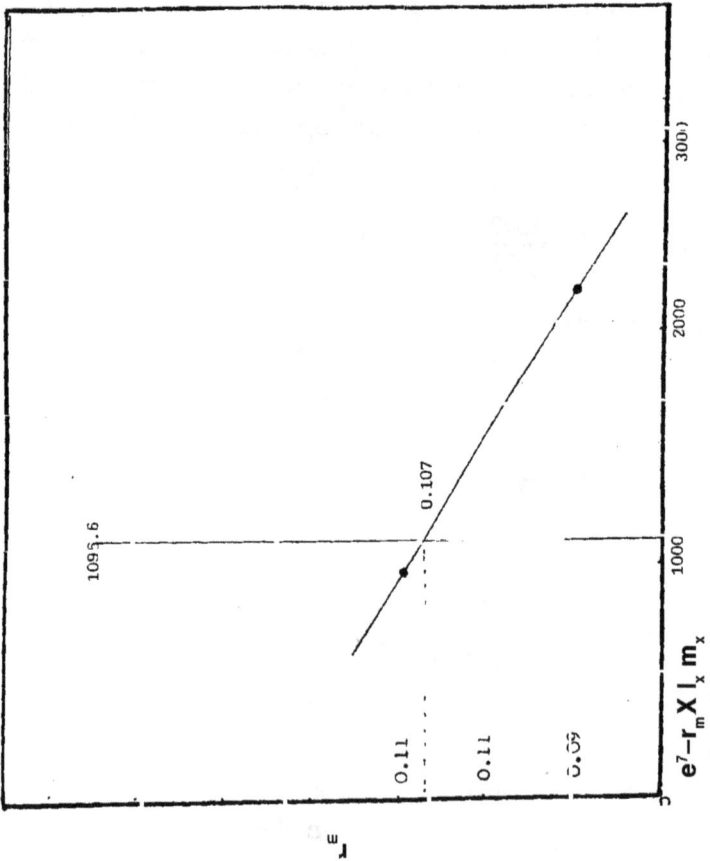

Figure 66: Determination of Intrinsic Rate of Increase in *A. lactinea*

Table 32: Female Progeny Production/Days by Mated Females of *A. lactinea*

Female No.	Number of Females Produced/Day					Total No. of Females Produced
	1	2	3	4	5	
A	36	21	8	6	0	71
B	28	18	5	0	D	51
C	46	23	19	14	D	102
D	61	46	22	18	7	154
E	57	39	19	13	D	128
F	56	43	25	13	D	137
G	43	20	23	11	D	106
H	47	27	11	D	—	85
I	37	18	6	D	—	61
J	27	20	14	9	D	70
Average	43.8	27.5	15.2	8.4	7	96.5

Table 33: Life Table Statistics of *A. lactinea*

Pivotal Age (Days) X	Proportional Life at Age x l_x	No. of Female Progeny/Female m_x	$l_x m_x$	$l_x m_x X$
42	1	43.8	43.8	1839.6
43	1	27.5	27.5	1182.5
44	1	15.2	15.2	668.8
45	0.8	8.4	6.72	302.4
46	0.2	0.7	0.14	6.44
47	0.0	0.0	0.0	00.00
			93.36	3999.74

Table 34: Provisional r_m (0.11) for *A. lactinea* and Related Values of $e^{7-r} m_x 1_x m_x$

x	$r_m X$	$7^{-r} m_x$	$e^{7-r} m_x$	$e^{7-r} m_x l_x m_x$
42	4.62	2.38	10.804	473.215
43	4.73	2.27	9.679	266.1725
44	4.84	2.16	8.671	131.799
45	4.95	2.05	7.767	52.194
46	5.06	1.94	6.958	0.974
				924.354

Table 35: Provisional r_m (0.09) for *A. lactinea* and Related Values of $e^{7-r} m_x 1_x m_x$

x	$r_m X$	$7^{-r} m_x$	$e^{7-r} m_x$	$e^{7-r} m_x l_x m_x$
42	3.78	3.22	25.028	1096.226
43	3.87	3.13	22.873	629.034
44	3.96	3.04	20.905	317.759
45	4.05	2-95	19.105	128.392
46	4.14	2.86	17.461	2.444
				2173.855

on the first day. The innate capacity for increase was found to be 0.190 (Figure 67) per female per day and population multiplied to 145.62 times in mean generation time of 26.21 days.

$$T_c = \frac{l_x m_x X}{l_x m_x} = \frac{3916.58}{145.62} = 26.8958$$

where T_c is arbitray T

$$r_c = \frac{\log_e R_o}{T_c} = \frac{\log_e 145.62}{26.89} = 0.185$$

where r_c is arbitray r_m and arbitrary 'r_m's (r_c) are 0.16 and 0.20

$$\therefore r_m = 0.190 \text{ (Figure 67)}$$

$$T = \frac{\log_e 145.62}{0.190} = 26.21 \text{ days}$$

Discussion

The intrinsic rate of natural increase (r_m) can be used for gaining insight into the population dynamics of a species. This information can be used as index of rate of population growth in a particular environment and potential effectiveness of a natural enemy (Messenger, 1964). The objective of the present study was to determine the intrinsic rate of increase *S. obliqua*, *A. lactinea* and *T. postica*. Several workers attempted such studies in different orders of the insect. Morris and Miller (1954) on *Choristoneura fumiferana* (Lepidoptera), Stark (1959) on *Recurvaria starki* (Lepidoptera), Richards and Waloff (1961) on *Phytodecta olivacea* (Coleoptera), Le Roux *et al.* (1963) on *Spilonota ocellava* (Lepidoptera), Waloff (1968) on *Sitona regansteinansis* Herbst (Coleoptera) and *Arytaina genistae*

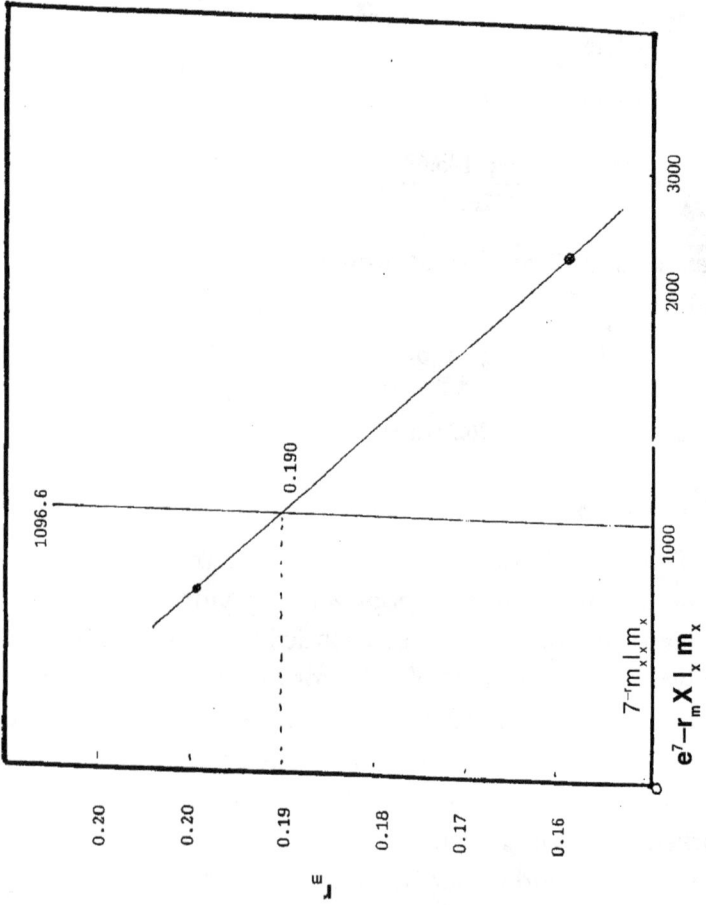

Figure 67: Determination of Intrinsic Rate of Increase in *T. postica*

Table 36: Daily Female Production by Mated Females of *T. postica*

Female No.	Number of Females Produced/Day					Total No. of Females Produced
	1	2	3	4	5	
A	57	43	34	D		134
B	66	44	34	9	D	153
C	49	47	41	15	D	143
D	43	42	38	25	D	135
E	62	53	34	D	–	149
F	77	56	27	0	D	162
G	62	54	31	14	D	161
H	58	47	40	D	–	145
I	61	48	33	D	–	142
J	55	46	22	24	D	147
Average	59.0	48.0	33.4	8.7	0	147.1

Table 37: Life Table Statistics of *T. postica*

Pivotal Age (Days) X	Proportional Life at Age x l_x	No. of Female Progeny/Female m_x	$l_x m_x$	$l_x m_x X$
26	1	5.9	59.00	1534.00
27	1	48	48.00	1296.00
28	1	33.4	33.40	935.20
29	1	8.7	5.22	151.38
30	0.6	0.0	0.0	00.00
			145.62	3916.58

Table 38: Provisional r_m (0.16) for *T. postica* and Related Values of $e^{7-r}m_x 1_x m_x$

x	$r_m X$	$7^{-r}m_x$	$e^{7-r}m_x$	$e^{7-r}m_x l_x m_x$
26	4.16	2.84	17.115	1009.785
27	4.32	2.68	14.5850	700.08
28	4.48	2.52	12.4285	415.095
29	4.64	2.36	10.59095	55.279
30	4.8	2.2	9.025013	00.000
				2180.239

Table 39: Provisional r_m (0.20) for *T. postica* and Related Values of $e^{7-r}m_x 1_x m_x$

x	$r_m X$	$7^{-r}m_x$	$e^{7-r}m_x$	$e^{7-r}m_x l_x m_x$
26	5.2	1.8	6.0496	356.89
27	5.4	1.6	4.9530	237.744
28	5.6	1.4	4.0551	135.440
29	5.8	1.2	3.3201	17.330
30	6.0	1.0	2.71828	00.000
				747.404

(Hemiptera), McLeod (1972) on *Neodiprion swainei* Midd. (Hymenoptera), etc.

In a sorghum stem borer, *Chilo partellus* (Swin.) Bains and Shukla (1976) reported that the intrinsic rate of increase (r_m) at different temperatures was in ascending order, 0.0002 (35°C), 0.165 (32.5°C), 0.223 (25°C), 0.383 (27.5°C) and 0.435 (30°C). The observations of Bains and Shukla (1976) shows that the rate of increase was maximum at 30°C which should be considered to be the optimum temperature for the multiplication of *C. partellus*. Further, Bains and Shukla (1978) reported weekly finite rates of increase at 25°C, 37°C, 32.5°C and 35°C were 4.67, 14.59, 21, 3.177, 2.002 respectively. While in the present study, 'r_m' was calculated for each pest species in respect of daily increase at laboratory conditions (25±1°C, 55–60 per cent RH, 12 hr photoperiod). In *Helicoverpa armigera* (Hubner) (Lepidoptera) Bilapate and Pawar (1980) studied the intrinsic rate of natural increase. They found that 285.06 females were produced per female during one generation and innate capacity and finite rate for increase in numbers were 0.1210 and 1.1280 respectively. The mean duration of a generation was found to be 46.71 days. The daily finite rate of increase of *H. armigera* was 1.1286 which enabled the insect to multiply 2.3322 times every week under the condition of abundant space.

Mani (1985) studied the age specific fecundity and rate of increase of *Eucelatorai brayani* Sabrosky on *H. armigera*. He found that the net reproductive rate (Ro) was 41.48 under caged conditions and with abundant host supply, the population of this tachinid fly increased at infinitesimal rate (r_m) of 0.1438 and finite rate of 1.1547 per female per day. The mean time for completing a generation was 25.9 days.

The life and fertility tables were constructed by Verma and Makhmoor (1988) for the cabbage aphid *Brevicoryne brassicae* under laboratory conditions on 6–7 weeks and cauliflower plants. The cohort had the maximum longevity of 33 days and the maximum period of reproduction was 15 days. The maximum average production of progeny was 5.75 nymphs per female per day on the 22nd and 25th day of life span. The species had the true generation time (T) of 21.711 days during which it multiplied 49.80 times (Ro). The intrinsic rate of natural increase (r_m) was 0.179 per female per day and the population doubled in 3.87 days. The finite rate of increase (λ) was 1.196 *i.e.* the species multiplied 1.196 times per day.

Nikam and Sathe (1983a) observed 0.176 the intake capacity increase per female per day in a hymenopterous parasitoid *Cotesia flavipes* (Cameron) and population multiplied to 30.72 times in mean generation time of 19.45 days. While, in *C. diurnii* Rao and Nikam the intrinsic rate of increase was 0.158 (Sathe, 1986).

Recently, Reddy and Bhattacharya (1988) studied the age specific survival mortality life-table and age specific-survival and fertility life table of *H. armigera* on four semi-synthetic diet. Various life parameters on these diets revealed that soaked form of soyabean based diet was highly suitable for the population growth of *H. armigera*. Different life parameters were calculated on this semi-synthetic diet and comparison was made to workout the suitability of population growth of *H. armigera*. The mean length of generation (T) of 32.4 days indicated that the insect would complete 11 generations in a year. The accurate and approximate estimate of intrinsic rate of increase (r_m) were 0.1008 and 0.1007 respectively. The finite rate of increase (λ) was 1.1061 per individual per day and the time required to double (DT) the population of *H. armigera* was 6.88. Net reproductive rate (R_o) doubling time (DT), potential fecundity (PF) and annual rate of increase (ARI) were 26.07, 6.88, 165.50 and 9.516 × 10^{15} respectively.

The present study was carried out at laboratory condition (25±1°C, 55–60 per cent, R.H., 12 hr photoperiod) and the intrinsic rates of increase were 0.163, 0.107, 0.190 and population multiplied to 174.42, 145.62, 93.36 times in 31.66, 26.21 and 42.39 days of mean generation time, in *S. obliqua, A. lactinea* and *T. postica* respectively. The 'm_x' maximum mean progeny production per day was 80.6, 43.8 and 59.0 on first day in *S. obliqua, A. lactinea* and *T. postica* respectively.

PARASITOIDS

Introduction

Biological insect pest suppression in its original or classical sense involves the directed use of beneficial organisms (Coppel and Mertins, 1977). These benificial organisms fall into several categories to include invertebrates such as parasitoids, predators, nematodes, pathogenic microorganisms like viruses, bacteria, fungi, rickettsiae, protozoans; vertebrate predators, such as birds, fishes, mammals, amphibians, etc. Most of these categories act as effective bio-control agent of insect pests. However, parasitoids and predators have immense value. The term parasitoid is different from parasite by representing following features in parasitoid.

1. The development of an individual destroies its host.

2. The host is usually of the same taxonomic class *i.e.* insecta.

3. Incomparison with their host, they are of relatively small size.

4. They are parasitic as larval and adults being free living.

5. Their action resembles with predators more than that of true parasites.

Parasitoids have been used more frequently than insect predators in bio-control programmes. Coppel and Mertin (1977) gave the ratio, parasitoids : predators, 3 : 1. The parasitoids have scattered in the orders, diptera and hymenoptera. The order hymenoptera possess complex fascinating biologies and frequently determine the pest population densities. They cause direct mortality towards the target pests. Thus, the parasitoids are of tremendous economic importance. The order diptera also contains sizeable number of parasitoids specially the family tachinidae is well known. Up-to-date more than 250000 species of parasitic hymenoptera have been estimated (Gupta, 1988). In past, Bhatnagar (1948), Rao (1961), Nixon (1965, 1967), Townes *et al.* (1961), Mason (1981), Sathe (1985a,b, 1986b, 1988a,b, 1990) Sathe *et al.*, (2003), Gupta (1976, 1987) have worked on insect parasitoids of pests.

Materials and Methods

For screening the parasitoids a large number of eggs and caterpillars of *S. obliqua*, *A. lactinea* and *T. postica* were collected from the different fields of Kolhapur. The field collected material was reared in the laboratory (25±1°C,

50-60 per cent R.H.,12 hr photoperiod) for parasitoid emergence. Observations were made daily on parasitoid emergence from the caterpillars. Later, per cent parasitism was calculated.

The cocoons of parasitoids were also collected from the field. Collected larvae were reared at the laboratory. All experiments were conducted in an insectary room (26±1°C, 50–60 per cent, R.H., 12 hr photoperiod). For the developmental studies fifty, 4 day old host larvae were exposed to five mated females in glass cages (Figures 3, 4) for parasitization. Parasitoid eggs and larvae were collected after 12 hr interval, dissecting parasitized host larvae in normal saline solution. Instars were identified observing size of head capsules and mandibles (Short, 1959, 1970). To determine the mating, newly emerged male and female pairs were caged in plastic containers (Figure 6) and copulation observed. The ovipositional behaviours were studied exposing 4 day old larvae to the mated females in glass cages (Figures 3, 4). The effect of different food on the adult longevity of parasitoid was tested in test tubes supplied with water/honey. The effect of host age on parasitization was studied by exposing hosts of different age with 50 host density for 24 hr, 100 per cent honey + water and fresh leaves of host plants were fed to the parasitoids and hosts respectively.

Results

S. obliqua

During the extensive rearing of *S. obliqua* caterpillars collected from field, a good number of insect parasitoids have been found emerged which are listed in Table 40.

Table 40: Parasitoids of *S. obliqua*

Sl.No.	Parasitoids	Stage of Attack	Order	Family
1.	*Apanteles obliquae* Wilkinson	L	Hymenoptera	Braconidae
2.	*A. creatonoti* Viereck	L	Hymenoptera	Braconidae
3.	*A. jayanagarensis* Bhatnagar	L	Hymenoptera	Braconidae
4.	*Cotesia flavipes* (Cameron)	L	Hymenoptera	Braconidae
5.	*Agathis* sp.	L	Hymenoptera	Braconidae
6.	*Parenion bhairavi* S. and I.	L	Hymenoptera	Braconidae
7.	*Balcemena* sp.	L	Hymenoptera	Braconidae
8.	*Agathis indica* Bhat	L	Hymenoptera	Braconidae
9.	*Foletesor rangini* S. and D.	L	Hymenoptera	Braconidae
10	*Habrobracon* sp.	L	Hymenoptera	Braconidae
11.	*Glyptapanteles malshri* S. and I.	L	Hymenoptera	Braconidae
12.	*Trichogramma evanescens minutum* Riley	E	Hymenoptera	Trichogrammatidae

Apanteles obliquae Wilkinson

A. *obliquae* is an internal larval parasitoid of *S. obliqua*. In the field 18 per cent parasitization was noted, it is potential parasitoid. The parasitoid completes its development from egg to adult within 18 days under laboratory conditions. Egg stage lasted for 3 days, 1st instar 4 days, 2nd instar 2 days, 3rd instar 3 days and pupa 7 days.

Longevity of the adult parasitoids have been studied. The females prolonged their live, maximum for 10 days (average 8 days) than males, 6 days (average, 5 days) with 100 per cent honey. Parasitoids die as quickly within 1 to 2 days in control reading. Mating takes place at day light within 12 hours, oviposition easily occurred on 2nd instars of *S. obliqua*, 2nd instar larvae were preferred for rapid parasitization. Laboratory reared parasitoids showed the sex ratio 1 : 1.25 favouring the females.

Apanteles creatonoti Viereck

The life cycle was completed within 16 days on *S. obliqua* larvae under laboratory condition. Egg stage lasted 3 days, larval stage 7 days and pupal stage 6 days. The females of *A. creatonoti* prolonged their live maximum for 13 days (average of 9.3 days) with 100 per cent honey solution.

As like many braconids, mating behaviour consisted attraction, recognition, orientation, wing fanning or vibrations, mounting, antenation, copulation and post copulatory grooming. Mating takes place within 40 seconds. 2nd instar host larvae were selected by the parasitoid preferably for parasitization. The sex ratio of parasitoids was 1 : 1.65, which favoured the females. 38 per cent mortality was caused by this parasitoid towards its host in fields.

Cotesia flavipes (Cameron)

C. flavipes is gregarious larval parasitoid of the sorghum stem borer. *Chilo partellus* (Swin.) but also recorded on *S. obliqua* in the field of Kolhapur. *C. flavipes* completed its life cycle within 17–18 days on *S. obliqua*. Egg hatching period was 2 days, 1st instar 3 days, 2nd instar 3 days, 3rd instar 3 days and pupa 6 days.

Three larval instars were found in *C. flavipes*, 1st two were vesiculate type and last was hymenoptriform. Longevity of both sexes ranged from 7 to 8 days with 50 per cent honey. Mating and oviposition easily occurred in the laboratory. Older instars of the *S. obliqua* were parasitized. Sex-ratio of the parasitoid (male : female) was 1 : 1.3.

Parenion bhairavi S. and I.

P. bhairavi is larval parasitoid and found potential biocontrol agent for *S. obliqua*. The detail studies of this parasitoid have been worked out recently by Inamadar (1990). The total life cycle from egg to adult was completed within 18 to 20 days. 10 per cent field mortality in larvae has been noted by this parasitoid.

Agathis indica

A. indica is larval parasitoid. In the field, 5 to 15 per cent parasitization was observed. The adult wasps are orange red in colour. The adults prolonged their live for 22 days with 100 per cent honey solution. Mating takes place soon after the emergence from cocoon and lasted for 1.5 minutes. Oviposition takes place on 2nd instar larvae.

Trichogramma evanescens minutum

The field collected eggs have screened the *T. evanescens*

minutum. It is minute egg parasitoid. 12–20 per cent mortality in eggs of *S. obliqua* has been noticed.

A. *lactinea*

For the survey studies of the parasitoids of *A. lactinea* eggs, larvae and pupae were collected from different fields of crops from Kolhapur. The laboratory reared material have shown the good number of hymenopterous parasitoids and dipterous parasitoids which are listed in Table 41.

A. *creatonoti*

The parasitoid attacked 2nd instar larval of *A. lactinea*. In the field 10 to 12 per cent parasitization was observed on the same pest.

A. *bosei*

This is important larval parasitoid of *A. lactinea*. Its parasitization reached upto 18 to 21 per cent in the field. The parasitoid completes its life cycle from egg to adult within 15 days. Egg stage lasted for 3 days, Larval stage 7 days and pupal stage 5 to 6 days.

The parasitoid is gregarious in habit. Both males and females survive for about 8 to 11 days under laboratory conditions with 100 per cent honey solution. Mating occured soon after the adult emergence from the cocoon. Oviposition is also easily accomplished within 2 seconds, during which a typical chain of behaviour was noticed; antennal examination of larvae, up-down movements of abdomen, ovipositor thursting, ovipositor insersion and actual oviposition in larva. Generally, 2nd and 3rd instar *A. lactinea* were preferred by the parasitoids for their egg laying. Laboratory reared parasitoids showed the sex-ratio (male : female), 1 : 1.1.

Table 41: Parasitoids of *A. lactinea*

Sl.No.	Parasitoids	Stage of Attack	Order	Family
1.	*Apanteles creatonoti* Viereck	L	Hymenoptera	Braconidae
2.	*A. bosai*	L	Hymenoptera	Braconidae
3.	*Glyptaponteles* sp.	L	Hymenoptera	Braconidae
4.	*Ecthromorpha* sp.	P	Hymenoptera	Ichneumonidae
5.	Tachinids	L	Diptera	Tachinidae

Glyptapaneteles sp.

Field parasitization by this parasitoid have been recorded upto 7 to 10 per cent. This is good complimentary, solitary, larval parasitoid of the above pest.

Ecthromorpha sp.

This is a pupal parasitoid of *A. lactinea*. 2 to 3 per cent parasitization was noticed in the field collected pupae. The parasitoid belongs to family Ichneumonidae. The female parasitized the pupal stages in the laboratory.

Tachinid Parasitoids

Three species of tachinid are screened from the field collected caterpillars. The tachinids attacked the older instars of the caterpillars, as much as 13 to 15 per cent parasitization was noticed in the field collected larvae. The parasitoids easily mate in the laboratory. With 100 per cent honey solution, the longevity was prolonged for 15 to 17 days.

Thiocidas postica Wlk.

T. postica is an important pest of ber tree, *Zizipus* spp. in Maharashtra, India. The caterpillars are only destructive since they feed on leaves and skeletonize the plant completely. During the survey of insect parasitoids a very large collection of the caterpillars were made. The laboratory rearing of these larvae showed four parasitoids which are listed in Table 42.

Discussion

The parasitoids attacks the immature stages of host insects. Some species (Euphorinae) reported to be attacking the adults (Clausen, 1940). Some are known attacking host

Table 42: Parasitoids of *T. postica*

Sl.No.	Parasitoids	Stage of Attack	Order	Family
1.	*Apanteles creatonoti* Viereck	L	Hymenoptera	Braconidae
2.	*Apanteles baoris* Wilkinson	L	Hymenoptera	Braconidae
3.	*Charops* sp.	L	Hymenoptera	Ichneumonidae
4.	*Tachina fallax* Meigen	L	Diptera	Techinidae
5.	*Enicospilus* sp.	L	Hymenoptera	Ichneumonidae

pupae (Beirsford *et al.*, 1970), host larvae (Broodryk, 1969) and eggs (Lewis and Redlinger, 1969).

In parasitic hymenoptera the number of larval instars are vary and differ from each other by cuticular ornamentation and size (body, head capsule, mandibles and spiracles). For distinguishing the various instars Thorpe (1930), Vance and Smith (1933), Cals and Shaumar (1965) etc. used the size of the mandible and head capsule. Minimum, single instar was reported by Arthur 1963 in *Exeristes constockii* (Cresson) (Ichneumonidae), Laing and Caltagirone (1969) in *Bracon lineatellae* (Fischer) (Braconidae) and Calvert and Bosch (1972) in *Monoctonus nervosus* (Haliday) (Braconidae) by observing the size of mandibles. The size of the spiracle opening was also used to distinguish larval instars in most species of parasitic hymenoptera. In the present study instars were identified with the help of increase in size of mandible and head capsule.

In most braconidae *Cotesia congregatus* (Say) a parasitoid of spingids (Fulton 1940), *Orgilus lepidus* Muesebeck (Oatman *et al.*, 1969), *Chelonus curvimaculatus* Cameron, a larval parasitoid of potato tuber moth, *Phthorimaea operculella* (Zeiler) (Broodryk, 1969), *Cotesia flavipes* (Cameron) (Kajita and Drake 1969) noted 3 instars. Similarly, in the present study 3 instars were noted in braconids. The 1st and 2nd instars were vesiculate and last was hymenopteriform. In *C. diurnii* (Sathe, 1986b) and *C. orientalis* (Sathe 1988b) the parasitoids of *Exelastis atomosa* Walsingham three instars were noted, of which, two instars were vesiculate and last was hymenopteriform. Again, in *Apanteles creatonoti* Viereck, Sathe *et al.* (1988b) reported three larval instars. However, Ichneumonids shows more

number of instars in their larval development as compared
to Braconids. Fisher (1959) reported 5 Instars in an
Ichneumonid, *Campoletis chlorideae* Uchida, a larval
parasitoid of *Helicoverpa armigera* (Hurn). Later, Tikar and
Thakre (1969) noted 4 instars in the same species. While,
Rojas-Rousse and Benoit (1977) reported 5 instars in *Pimpla
investigator* (F), a common endoparasitoid of lepidopterous
pupae. Likewise, Sathe and Nikam (1986) also reported five
instars in another Ichneumonid wasp, *Diadegma trichoptilus*
(Cameron), a solitary, larval parasitoid of *E. atomosa*. Very
recently again, Sathe (1990) pointed out 5 instars in
D. argenteopilosa (Cameron) (Ichneumonidae) in which first
two instars were caudate type and other three were
hymenopteriform.

In a chalcid parasitoid, *Brachymeria lasus* (Walker),
(Chalcidae) Narendran and Joseph (1976) reported 5 instars
out of which 1st instar was hymenopteriform. But in
B. compsilurae (Dowdon, 1938), *B. fonseolombei* (Parkur
1924), and *B. minuta* (Sychevskaya, 1966) the 1st instar larva
was caudate type instead of hymenopteriform. However,
the total number of instar was five in all the species.

Short life cycle leads fast rearing and hence it is one of
the desirable attribute of an ideal parasitoid. In *C.
congregatus,* a parasitoid of the tomato worm, *Protoparce
sexta* (Johan), the duration of development from egg to adult
was 19–20 days (Fulton, 1940). In *C. curvimaculatus* the egg
hatching period was 1–1.5 days, 1st instar lasted 7 days, 2nd
instar 2–3 days, 3rd instar 0.5–2 days, prepupa 1–1.5 days
and pupa 5–6 days thus, the minimum and maximum
duration of development from egg to adult was 16.5 days
and 21 days respectively (Broodryk, 1969) while, in

D. argeneopilosa (Sathe, 1990) the total developmental period was 18 days.

The studies were conducted on *Cotesia margniventris* (Cresson) by Jalali *et al.* (1987) under laboratory condition wherein the oviposition was preferred on 3–5 days old larvae, the developmental time from oviposition to cocoon formation lasted 8.2±1 days and pupae period 3–4 days. In *Praon peguoerum* Viereck (Hymenoptera : Aphidiinae), a parasitoid of pea aphid the adult emergence required 13 days. While, in a Chalcid parasitoid, *B. lasus* the minimum duration was 10 days and maximum 18 days for the development from egg to adult (Narendran and Joseph, 1976).

Since food is population regulatory abiotic factor of insects in nature and in laboratory the studies were made on parasitoids of the pests considered here. In past, Kajita and Drake, (1969), Cardona and Oatman (1971), Odebiyi and Oatman (1972), and Sathe and Nikam (1983a, 1985) have searched for the nutritional requirement in parasitoids.

Kajita and Drake (1969) found that the longevity of *C. chilonis* was decreased when fed with water and increased when diluted honey was supplied. Oatman *et al.* (1969) reported that the longevity of both sexes of *O. lipidus* was considerably greater with water and honey than that when either was provided alone. The males of *A. dignus* lived longer than females; without food or water longevity decreased, remarkably greater longevity of both the sexes was with honey and when provided with fresh leaflets, the adults died quickly as like control (Cardona and Oatman, 1971). Odebiyi and Oatman (1972) also reported similar observations in *A. gibbosa*. But in *Microbracon chelonis* Viereck, a parasitoid of *Chilo zonellus* (Swin.) the longevity was increased with sugar solution. In the present forms, the longevity was found increased considerably with honey solution.

Mating has direct influence on biocontrol programme, since unfertilised eggs produced only males in parasitic hymenoptera. The mating behaviour sequence typically consisted a chain, attraction recognization, orientation, wing fanning, or vibrations, mounting, antennation, copulation and post copulatary groming (Cole, 1970) in most of the braconids and Ichneumonids. Broodryk (1969) and Cardona and Oatman (1971) noted polygamous males in *C. curvimaculatus* and *A. dignus* while, females were monogamous. But, Gordh and Hendrickson (1976) reported polygamous phenomenon in both sexes of *Bathepletis curculionis* (Thompson), a parasitoid of *Hypera postica*. In *C. orientalis* no multiple matings were observed (Sathe and Nikam, 1984) while in *C. diurnii* and *A. prodeniae* males were polygamous and females monogamus (Sathe, 1985b; Santhakumar, 1989).

Weseloh (1977) reported that the females of *C. melanoscelus* were receptive to males for first 11 days of their adult life and were coarted. The females of *A. prodeniae* were receiptive for 3 days only after emergence while, the females of *C. diurnii* were receptive for 2 days (Santhakumar, 1989). In the present study, the females were receptive for 3 days.

In *Priopoda nigricollis* (Thoms), a parasitoid of Birch leaf miner, *Fenusa pusilla* the copulation was lasted from 3" to 7" (Quendnau and Guevremont, 1975), in *B. cuculionis* it was 3.58'–16' (Dowell and Horn, 1975) whereas, in *A. creatonoti* the same was lasted for 40".

Habitat selection, host location within the selected habitat, acceptance (or rejection) of host and ability (or enability) of the parasitoid to grow satisfactorily on the host's tissues and the important aspects of parasitoid attacking behaviour (Smilowitz and Iwantsch, 1972) in parasitic hymenoptera (Ayyar and Narayannaswami, 1940; Nishida. 1956; Loan, 1965; Leong and Oatman, 1968; Broodryk, 1969; Oatman *et al.,* 1969; Cardona and Oatman, 1971; Odebiyi and Oatman 1972; Quednau and Guevermont, 1975; Sathe and Nikam, 1983b–84) have worked out widely.

In *A. gibbosa* the time required for egg deposition was 30" (Odebiyi and Oatman, 1972). The periods required for egg deposition by *C. diurnii, C. chlorideae, E. argenteopilosus* and *A. prodeniae* averaged 3", 4", 2" and 4" respectively (Santhakumar, 1989). In present parasitoids egg deposition was accomplished within 2" in *A. obliquae* and *A. creatonoti.*

In general, five successive phases of oviposition behaviour were displayed by the present parasitoids *viz.* antennal examination of larva, up-down movements of abdomen, ovipositor trusting, ovipositor insertion and actual oviposition towards their respective hosts as many parasitoids do (Odebiyi and Oatman, 1972). The whole oviposition process was completed in few seconds. Host instars and host species preference of parasitoids are useful aspects for introduction, colonization and mass production of parasitoids.

Trichogramma evanescens Westwood preferred younger eggs of *Cadra cautella* (Lepidoptera) for maximum parasitism (Lewis and Redinger, 1969). Similarly, the female of *C. curvimaculatus* also preferred the younger eggs for maximum parasitism (Broodryk, 1960) while, the larval parasitoid *O. lepidus* preferred 3–4 days old larvae most for parasitization and a maximum of 92.2 per cent adult parasitoids were emerged (Oatman *et al.*, 1969). In *A. dignus* and *A. gibbosa* the optimum age for maximum parasitization was 2–3 days in both the cases (Cardona and Oatman, 1971; Odebiyi and Oatman, 1972). However, the larvae of 8–9 days were not suitable to progeny production.

Smilowitz and Iwantsch (1975) reported maximum, 31 per cent parasitism on 1st instars of its hosts and further they found that declined in older instars, while in *C. congregatus* the development of parasitoid was slower, when oviposition occurred in 1st, 2nd instars of *Manduca sexta* (L) than occurred in 3rd and 4th instars (Thurston an Postley. 1968); Hopper and King (1984) found the host instar preferences in *Microplitis croceipes* (Braconidae), a parasitoid of *Zea boddies* and *H. virescens* (Fabricius) in which parasitoids preferred 3rd instar larvae most, *C. flavipes* preferred older instars for maximum parasitism (Nikam and Sathe, 1983b), while the recent reports of Nair (1988) showed that the wasp, *C. flavipes* successfully parasitized the diapausing host larvae. The present data will serve as basis for mass culturing programme of parasitoid species considered under this chapter.

PREDATORS

Introduction

Almost all animal and plant species have natural enemies which cause temporary or minor effect or the death of host. Predators are important among them. The predators have distinguishing characteristics *viz.* it generally consumes more than 1 host individuals. Most predators move around freely in both their immature and adult stages, while searching for and feeding on their prey. Generally, predators are larger than their prey. About 167 families of 14 orders contain predatory insects. However, the orders Coleopetera, Neuroptera, Hymenoptera. Diptera and Hemiptera provide a very large number of individuals which frequently control the pest populations. The first dramatic example of deliberate manipulation of insect natural enemies was the importation of the Vedalia lady bettle, *Rodolia cardinalis* (Mulsant) into California in 1888 to control the cottony cushion scale *Icerya purchasi* Maskell on citrus. It is estimated that possibly up to one-third of the successful biological insect pest suppression programmes are attributable to the introduction and release of insect predators (Sweetman, 1958). Most predators directly involved in these programmes are coccinellids but, carabids, chrysopids, formicids, syrphids and mirids have also been used. Most preys suppressed successfully are homopterans. Whitecomb (1973) illustrated our lack of knowledge of predacious arthropods and discussed the importance of certain groups of predators. Many predacious species from cruciferous crops he mentioned in the list (Root, 1973). Earlier workers on prey predator relationships are Agarwal *et al.* (1983), Anonymous (1985, 1986), Bakhetia and Sidhu (1977), Butani and Bharodia (1984), Bose and Ray (1967), Chakravarthy and Lingappa (1979), Singh and Singh (1983), Jotwani *et al.* (1961), Kamat *et al.* (1970), Kumar and Ananthakrishnan (1984), Kulshrestha and Agarawal (1982), Kamath (1960), Patel and Vyas (1984), Rao (1980), Rajendra and Patel (1971), Rajasekhara *et al.* (1964), Rajasekhara and Chatterji (1970), Sethi *et al.* (1976), Singh and Gangrade (1974, 1975), Sharma (1975), Talati and Butani (1979), Thobbi and Singh (1974), Singh (1975), etc.

In India our knowledge on the predacious insect is mainly restricted on the study carried out by CIBC Indian Station (Rao, 1969) though some scattered reports and the work of Sathe and Bosale (2001), Patil and Sathe (2003) and Sathe and Shinde (2008).

Materials and Methods

Feeding behaviour and Predatory prey relationship was studied by caging the predators. *Rhynocoris fuscipes* (F.), *Cantheconidea furcellata* (Wolff), *Scadra annulipes* Reut. and *Andrallus spinidens* (F), individually with *S. obliqua* early instar larvae in glass cages (Figures 3, 4). Observations were made on their feeding behaviour. Mortality rate per day of the *S. obliqua* larvae due to predators was studied with respect to different host densities, exposing in glass cages for 24 hr. Prey consumption rate by the predator species were studied by exposing optimum prey density till the death of predator species. The culture of predatory insects were started by collecting nymphal and adult stages of the predatory insects along with the prey insects from fields. *Spilosoma* larvae used in these experiments were maintained by providing them sunflower leaves as a food.

Results

Observations on Predator's Feeding Behaviour

Almost all species of the predatory insects showed a series of steps when provided prey for their feeding. The following sequence of the behaviours of predator have been noticed:

1. Distance reaction to prey.
2. Prey recognition.
3. Up and down movements of second pair of legs.

4. Cleaning of mouthparts.

5. Pause.

6. Orientation (positioning of predator behind the prey).

7. Pouncing of prey (from behind).

8. Clutching of thoracic region of prey.

9. Immediate paralysing of prey or struggling of prey.

10. Dorsoventral twisting of prey.

11. Circling of predator with prey.

12. Puncturing of sternite of prey.

13. Suction of inner contents.

14. Prey rejection (Prey becomes shapeless mass).

15. Pause

Mortality in Relation to Prey Density

The predators prefer the young larvae of *S. obliqua*. On an average *R. fuscipes* kills 3 larvae per day. The prey prolonged its live on an average 25.8 days and consume 79 prey larvae. Other species found comparatively less effective (Tables 43 and 44). The order of prey mortality due to predator species was *C. furcellata* > *S. annulipes* > *A. spinidens*.

Consumption Rate of Prey

55, 39 and 36 caterpillars of *S. obliqua* were consumed by *C. furcellata*, *S. annulipes* and *A. spinidens* (Table 44), during their average life-span of 22, 19 and 24 days respectively. Per day rate of consumption was also less in the other species compared to *R. fucipes*.

Table 43: Mortality Due to Predators in Relation to Different Prey (*S. obliqua*) Density

Predator Species	Average Prey Density					
	3	5	10	15	20	25
R. fusipes	2.0	2.3	3.0	3.0	3.0	3.0
C. furcellata	1.5	2.0	2.3	2.5	3.0	3.0
S. annulipes	1.0	2.0	2.0	2.0	2.0	2.0
A. spinidens	1.5	1.5	1.5	1.5	1.2	1.2

Table 44: Consumption Rate of Prey During Life Span of the Prey Species

Predator Species	Prey Density Exposed/Day	Total Number of Prey Consumed	Prey Longevity
R. fuscipes	10	79.00	25.8
C. furcellata	10	55.00	22.0
S. annulipes	10	39.00	19.0
A. spinidens	10	36.00	24.0

Discussion

Vijayalaxmi (1986) made some observations in laboratory to understand the mode of attack of *Heteropoda venatoria* (L), approach towards the prey and ability to regulate the population of prey *Peripianata amiericana*. In this case, predatory response was governed by prey size and density. Experimental results showed that larger predators do not attack the smaller prey beyond a particular size, hence competition for prey with the smaller size, predators avoided by the optimum prey size selected system.

Ghosh and Chakrabarti (1986) reported 20 species of predators comprising of spiders and members of insect orders Coleoptera, Diptera, Heteroptera, Neuroptera associated with 15 aphid species. Maximum predatory activity was noticed during the winter months when most of the prey aphid species reproduce profusely.

Bakhetia and Sidhu (1977) studied the biology, seasonal occurrence, food plants and natural enemies of *Aphis craccivora* Koch, a major pest of groundnut. They found the predators *Monochilus saxmaculatus* (F), *Brumoides suturalis* (F.) and *Coccinella septumpunctata* L. breeding and feeding on the aphid in the field.

Hedylepta indicata (F) (*Lamprosoma indicata*) and L. *diemenalis* (Gn.) sporadically cause considerable damage to crops of soyabean in India (Bhattacherjee, 1977). Their outbreaks occur only sporadically, because the pests were held in-check by natural enemies, predatory spiders were found infecting pest in September–November.

Bose and Ray (1967) made comparative study on the consumption of aphids by the common predator *Chilomanes saxmaculata* Fabr. (Coleoptera : Coccinellidae). They showed that consumption increased daily upto the fifth day after hatching and thereafter decreased as the larvae prepare to pupate. A single larva consumed 143–189 individuals of *A. narii* or 355–394 of *A. craccivara* during its development, adult females were more voracious than males or larvae.

Butani and Bharodia (1984) studied the relationship between *A. craccivora* and its predator on groundnut in Gujarat, India, in 1981. The predators they considered were identified as *Hippodamia variegata* (Goeze) *Coccinella septempunctata* L. and *Menochilus sexmaculatus* (F.). An infestation index was calculated by dividing the number of plants examined. They observed positive correlation between the aphid index and the population of active stages of the predators during March, while in April the aphid population decreased with increased abundance of the coccinellids. The predator population had declined by the 1st week of May. Most of them have migrated to other area due to the decrease in prey abundance. They have concluded that application of insecticides is unnecessary if coccinellids are present in the groundnut crop in summer.

While searching for the natural enemies of *S. obliqua*, Chand and Prasad (1970) found nymphs and adults of *C. furcellata* (Wolff) preying on young larvae of *S. obliqua* in the field and this was the first record of the pentatomidae attacking the pest.

Kapoor *et al.* (1975) studied the predator prey relationship of *C. furcellata* and *Prodenia litura* Fab. Their laboratory studies showed that the predators egg stage and 5 nymphal instars lasted 5–7, 2, 3, 2–3, 2–4 days respectively. Nymphs consumed a total of 14–19 larvae during their development and a pair of adults 48–68 individuals in 10–12 days. While in present study the same predator found consuming maximum 79 young caterpillars of *S. obliqua*.

Khan and Sharma (1972) observed for the first time attacking and carrying away the larvae of *Helicoverpa armigera* (Hubn.), *Mythimna separata* (Wlk.) *Agrotis segetum* (Schiff.) and other species of *Agrotis* in field crops in Rajasthan, India by the foraging workers of the ant, *Cataglyphis bicolor* (F.). Further they noted, substantial reduction in the population due to the predation of this ant.

Kumar and Ananthakrishanan (1984) studied the predator-thrips interaction with reference to *Orius maxidentex* Chauri and *Carayenecoris indicus* Muraleedharan (Anthocoridae : Heteroptera). They studied the anthocorids, *O. maxidentex* and *C. indicus* in the laboratory and the field near Madras, India, as predators of thrips *O. maxidentex* was present from January on sesume, it fed on *Thrips palmi* Karny on the young foliage, migrated later to prey on *Frankliniella schultzei* (Tryb.) on the flowers.

Larvae and adults of *Microlestas discoidalis* (Fairm.) were found preying on larvae of *Agrotis* (Euxoa) *segetum* (Schiff.) at Jaipur, Rajasthan, India, where the Noctuid was an important pest of pea crops (Mathur *et al.,* 1971). In the laboratory, larvae of the carabid in the course of their development destroyed upto 32 larvae of the Noctuid in 13 days and adult beetles destroyed several hundred larvae. They also reported that *Anthicus crinitus* Laforte was seen in association with the carabid, feeding on injured larvae of *Agrotis*.

Pawar and Jadhav (1983) reported that in field and cage studies the wasps, *Dalta pyriformis* F., *D. carnpaniforme* (F.), *D. esuriens* (F.) and *D. conoideum* (Gmal.) were found preying on larvae of *H. armigera* on pigeonpea.

During collection, *Orius tantilus* (Motsch) found preying on *Taeniothrips nigricornis* (Schmutz) infesting the flowers of *Cajanus cajan* at New Delhi, another bug was seen preying on the thrips (Rajasekhara, 1964). This was identified as *Psallus* sp. Laboratory studies showed that it completed its development from egg to adult in 20–21 days and fed on the nymphs and adults of *T. nigricornis*. Rajendra and Patel (1971) studied the life-history of a predatory Pentatomid bug, *A. spinidens* on *H. armigera*, wherein they found that females laid 11–1084 eggs each with an average of 370. The incubation period was 5.29±0.96 days at room temperature fluctuating between 85 and 92°F and 7.63±0.51 days at a constant laboratory temperature of 80±2°F. The hatching rate decreased with decreasing humidity from 95 per cent at 100 per cent RH to 0 per cent at 20 per cent RH. The nymphal stage lasted, 12.48±0.5 days at an average temperature of 87.09±3.89°F and 21.98±1.78 days at a constant temperature of 80±2°F.

The syrphid *Sphacrophoria scutellaris* F. is of potential importance in India for the control of aphids, particularly *Rhopalosiphum pseudobrassicae* (Davis) on which the larvae feed and the effect of nutrition on its length of life and fecundity was studied under controlled conditions of temperature and humidity by Lal and Haque (1955). They showed that the larvae completed their development in 12–13 days and destroyed 401–493 aphids each at a temperature of 19.8°C and 68.3 per cent relative humidity and completed their development in 8–9 days and consumed 308–339 aphids each at 22.2°C and 71 per cent RH. Experiments at 20, 25 and 30°C and 50, 70, 90 per cent RH showed that both factors had an important effect on the survival and fecundity of the adults, which were significantly greater, at 20 than 25 or 30°C and 70 and 90 than at 50 per cent RH. At 25°C the insects lived longer at 90 per cent RH that at 70 or 50 per cent; whereas at 20°C they lived longer at 70 per cent RH than at 90 per cent. In tests on adult nutrition, sucrose, honey, fructose, glucose, maltose and mannitol were more effective, in promoting longevity than starch, yeast or water. Of the sugars, sucrose caused the greatest increase in length of life and fecundity.

Artificial rearing of the predator should, therefore, be carried out at a temperature of 68°F and 70 per cent RH, and sucrose should be provided for the adults.

Thobbi and Singh (1974) reported for the first time the role of *H. armigera* as a predator of the pupae of castor semilopper, *Achaea janata* (L). Vasantharaj and Janagarajan (1966) reported a new host, *Dactynotus carthami* H.R.L. for the predatory coccinellid *Brumus suturalis* Fab. They reported that 1 adult coccinellid can consume about 11 aphids in a day.

Singh and Gangarde (1974) while studying the biology of *S. obliqua* found a reduviid, *R. fuscipes* preying on young larvae of *S. obliqua*. Again, Singh and Gangarde (1975) reported the predator, *R. fuscipes* killing 1–3 larvae daily during its life-span of 21–29 days, while in the present study, the same pentatomid consumed maximum 79 larvae of 1st instar of *S. obliqua*. Other predators tried against *S. obliqua* were comparatively less effective.

DISEASES

Introduction

Insects are attacked by bacteria, fungi, viruses, protozoans, rickettsia and nematodes. Some of these pathogens causes eipzootics in natural insect populations, quite commonly and frequently, but some pathogens may occur infrequently and some of the pathogens cause high mortality towards its host but, others may produce only cronic effects. Insect pathogens or microbials act naturally to limit populations of crop pests as like other natural control factors. Micro-biocontrol is defined by Falcon in 1971 as "Including all aspects of the utilization of microorganisms or the bi-products in the control of insect pest species". Micro-biocontrol also be defined as "that part of biological control concerned with controlling insects by the use of microorganisms (including viruses)". More than 2000 insect pathogens have been described (CSIR Author, 1980). Pathogens exert their controlling effect by means of their invasive properties by toxins, enzymes and other substances. This is probably a small fraction of the total number of pathogens affecting insects however, little is known about many of these pathogens but, some pathogens have been studied extensively. As a fact, insect pathogens can be used (*i*) by utilizing natural occurring diseases, (*ii*) by the introduction of insect pathogens to the pest population as permanent mortality factor and (*iii*) by the repeated applications of microbials for temporary pest control.

Microbials have many characteristics ideally suited for use in pest management programme. They are usually specific and highly virulent on given host, pose little hazard to non-target organisms and are usually comparable with other management programmes. Although some degree of success has been achieved with their use in the pest management. However, at present microbials are used at a very little quantity compare to chemical pesticides. Diseases are not used due to the unpotentially of microbials on large scale but, the lack of understanding its identity and effective use in pest management. Keeping in view the above facts the present topics is represented.

In past, Bell (1981), Bell and Romine (1980), Falcon (1974), Mech (1974). Mohanad *et al.* (1978), Ignoffo and Hostetter (1984), Oatman *et al.* (1970) and others studied the pest control with respect to pathogens. The microbial control has practical potential. In most field trials the virus and bacteria appear to have been more consistent in their effect hence, it is extremely essential to enlight and popularise the microbial measure in pest management. For successful implementation of this technique, survey, identification. abundance, mortality rate, etc. are worthwhile aspects. As a most promising bacterium available for microbial control of insects is *Bacillus thuringiensis* Berliner, has been widely used against several harmful lepidoptera in many countries.

Materials and Methods

Caterpillars of *Spilosoma obliqua* (Wlk.) and *Amsacta lactinea* (Cramer) were collected from different fields of sunflower during August to October 1988–1990. The field collected materials were reared in the laboratory (25±1°C, 55–60 per cent R.H., 12 hr photoperiod) for screening their various pathogenic biocontrol agents. From the diseased caterpillars, pathogenic organisms were isolated and preserved.

Bacteria Isolation

The first step in isolating bacteria from a diseased insect was "sterilization of specimen to remove contaminated saprophytic forms. Later, the specimen was dipped into 70 per cent or 95 per cent ethanol (wetting agent) for 2 seconds and transferred to a solution of sodium hypochlorite for 3.5 minutes, then placed for an equal period in 10 per cent sodium thiosulfate to remove free chlorine. After the external surface has been sterilized, the specimen was rinsed in three changes of sterile distilled water and placed on a sterile dissecting dish. The specimen was then opened with sterile scissors by cutting the integument along a longitudinal dorsal line, with care being taken not to cut into the gut eipthelium. Blood and body fluids were sampled with a sterile capillary tube, diluted in 2 ml sterile Ringer's solution, and then placed on NA, BHIA or other suitable bacterial media by the streak plate method. Tissue were examined by removing a small sample, placing it in 1 ml of sterile Ringer's and then triturating it with a sterile glass rod. The suspension was then infusion streaked on a place of nutrient or brain-heart infusion agar. The agar plates were inverted and incubated at room temperature or 30°C overnight and then examined for well-isolated colonies. These colonies represented pure bacterial strains and were identified.

Fungi Isolation

The following steps were adopted for isolation of fungi from diseased insects:

1. Surface sterilisation of the insect by immersing it in a 5 per cent solution of NaClO for several minutes, then rinsed it in three changes of sterile water.

2. In a sterile dish, the specimen was opened and transferred a small portion of infected tissue to a sterile culture plate.

3. The cultures were placed in a moist incubator at 25°C and examined.

Virus Isolation

The diseased insects were placed in a culture tube with water. After two or three days, the inclusion bodies were accumulated as a white layer on the bottom of the tube. Cell and tissue remnants, bacterial cells and other breakdown products were separated from the inclusion bodies by repeated washing and differential centrifugation. If the viruses are to be used for infection studies, the pellets are treated with antibiotics for 24 hr and washed several times with sterile distilled water and differentiated by centrifugation, generally, removed most bacteria. A variation which was particularly helpful in purifying very small inclusion bodies, such as granulosis capsules and small cytoplasmic polyhedra, was mixed with triburated diseased insect with an equal volume of carbon tetrachloride shaked vigorously and centrifuged at 3000 rpm for about 1 hr. A layered plug of inclusion bodies formed between the CCl_4 (bottom) and the water phase while, fat accumulated on top of the water phase. The water and fat were decanted off without disturbing the plug and the inclusion bodies which formed the top layer of the plug, were carefully washed with a small spatula and suspended in water. Further, purification was accomplished by repeated washing and differential centrifugation.

Experiments were also carried out with respect to mortality caused by granulosis virus 10^7–10^9 spray on *S. obliqua, A. lactinea* and *T. positca*. In a glass trough, caterpillars were confined with different host densities along with their host food plant leaves. The food materials were sprayed with granulosis virus 10^7–10^9 and mortality in 4th instar caterpillars were noted.

Resutls

S. obliqua

During the screening of pathogenic diseases of *S. obliqua*, following fungal species have been isolated from the field cultures.

1. *Aspergillus flavus*

2. *Nomuroea rileyi*

3. *Entomophthora* sp.

4. *Entomophthora virulenta*

Out of above fungi species, *Aspergillus* sp. and *Entomophthora* sp. were predominant in their appearance on *S. obliqua*. During the survey studies (1988–1990) maximum, 9.8 per cent mortality of *S. obliqua* was noted due to *Aspergillus* sp. and *Entomopthora* sp. (Tables 45–47). The results on mortality due to fungi are tabulated in Tables 45 to 47. As regards Bacteria, *Bacillus thuringiensis* was the only species found attacking the caterpillars of *S. obliqua*. Maximum 20.31 per cent mortality was noted by this species. The results are recorded in Tables 48–50.

As regards to viruses, two types of viruses have been reported on *S. obliqua* from Kolhapur region. Maximum mortality due to viruses was 20.61 per cent in the caterpillars of *S. obliqua*. The results are shown in Tables 51–53. The laboratory experiments on granulosis virus spray showed maximum 60 per cent mortality in 4th instar larvae of *S. obliqua*. The results are represented in Table 54.

A. lactinea

A. lactinea caterpillars were attacked by a fungus, *Aspergillus funigatus*. Nuclear polyhedrosis virus (NPV). Maximum, 32.75 per cent caterpillars which collected from fields were attacked by fungus and minimum 5.55 per cent (Tables 55–57). The minimum and maximum percentage of mortality due to NPV virus was 28.16 per cent and 6.89 per cent respectively (Tables 58–60). In the laboratory experiments maximum 60 per cent and minimum 8 per cent mortality was caused due to the application of granulosis virus spray (Table 54).

Table 45: Mortality in *S. obliqua* Due to Fungi

Month	Date	Number of Larvae Collected	Aspergillus sp.		Entomopthora sp.	
			Mortality	Per cent Mortality	Mortality	Per cent Mortality
August 1988	6	118	4	3.38	5	4.23
	13	115	7	6.08	3	2.60
	20	1 21	2	1.65	3	2.47
	27	120	8	6.66	6	5.00
September 1988	3	1.3	9	8.73	6	5.8
	10	104	10	9.61	2	1.92
	17	96	7	7.29	8	8.33
	24	109	2	1.83	10	9.90
October 1988	1	92	4	4.34	8	8.69
	8	130	9	6.92	10	7.69
	15	91	8	8.79	3	3.29
	22	84	5	5.95	6	7.14
	29	71	2	2.81	3	4.22

Table 46: Mortality in S. obliqua Due to Fungi

Month	Date	Number of Larvae Collected	Aspergillus sp.		Entomopthora sp.	
			Mortality	Per cent Mortality	Mortality	Per cent Mortality
August 1989	5	0	–	–	–	–
	12	0	–	–	–	–
	19	22	–	–	1	4.54
	26	31	1	3.22	1	3.22
September 1989	2	46	2	4.34	2	4.34
	9	59	4	6.77	2	3.38
	16	68	4	5.88	5	7.35
	23	73	6	8.21	7	9.58
	30	89	7	7.86	7	7.86
October 1989	7	103	10	9.70	9	8.73
	14	102	9	8.82	8	7.84
	21	101	8	7.92	8	7.92
	28	90	8	8.88	6	6.66

Table 47: Mortality in S. obliqua Due to Fungi

Month	Date	Number of Larvae Collected	Aspergillus sp.		Entomopthora sp.	
			Mortality	Per cent Mortality	Mortality	Per cent Mortality
August 1990	4	0	–	–	–	–
	11	0	–	–	–	–
	18	0	–	–	–	–
	25	21	–	–	1	4.76
September 1990	1	26	–	–	1	3.84
	8	28	1	3.57	1	3.57
	15	32	1	3.12	1	3.12
	22	48	2	4.16	2	4.16
	29	56	3	5.35	2	3.57
October 1990	6	104	9	8.65	6	5.76
	13	98	9	9.18	6	6.12
	20	102	10	9.80	8	7.84
	27	112	11	9.82	11	9.82

Table 48: Mortality in S. obliqua Due to B. thuringiensis

Month	Date	No. of Larvae Collected	Bacillus thuringiensis	
			Mortality	Per cent Mortality
August 1988	6	–	–	–
	13	–	–	–
	20	40	3	7.5
	27	59	5	8.47
September 1988	3	61	7	11.47
	10	71	8	11.26
	17	89	3	3.37
	24	98	10	10.20
October 1988	1	104	4	3.84
	8	114	21	18.42
	15	116	6	5.17
	22	119	23	19.32
	29	107	21	19.52

Table 49: Mortality of S. obliqua Due to Bacillus thuringiensis

Month	Date	No. of Larvae Collected	Bacillus thuringiensis	
			Mortality	Per cent Mortality
August 1989	5	–	–	–
	12	12	–	–
	19	22	1	4.54
	26	39	2	5.12
September 1989	2	41	3	7.31
	9	59	2	3.38
	16	58	5	8.62
	23	71	8	11.26
	30	74	9	12.16
October 1989	7	79	4	5.09
	14	88	16	18.18
	21	109	21	19.26
	28	128	26	20.31

Table 50: Mortality of *S. obliqua* Due to *Bacillus thuringiensis*

Month	Date	No. of Larvae Collected	Bacillus thuringiensis	
			Mortality	Per cent Mortality
August 1990	4	–	–	–
	11	16	1	6.25
	18	26	2	7.69
	25	39	4	10.25
September 1990	1	41	4	9.75
	8	51	6	11.76
	15	61	9	14.75
	22	59	11	18.64
	29	78	15	19.23
October 1990	6	109	18	16.51
	13	116	7	6.03
	20	115	3	2.60
	27	119	5	4.20

Table 51: Mortality in S. obliqua Due to Viruses

Month	Date	No. of Larvae Collected	Viruses	
			Mortality	Per cent Mortality
August 1988	6	–	–	–
	13	–	–	–
	20	39	–	–
	27	41	5	12.19
September 1988	3	53	7	13.20
	10	52	3	5.76
	17	69	13	18.84
	24	78	11	14.10
October 1988	1	108	17	15.74
	8	111	22	19.81
	15	98	20	20.40
	22	104	4	3.84
	29	100	20	20.00

Table 52: Mortality in *S. obliqua* Due to Viruses

Month	Date	No. of Larvae Collected	Viruses	
			Mortality	Per cent Mortality
August 1989	5	–	–	–
	12	–	–	–
	19	–	–	–
	26	39	–	–
September 1989	2	43	4	9.30
	9	42	5	11.90
	16	61	8	13.11
	23	79	13	16.45
	30	83	14	16.86
October 1989	7	97	20	20.61
	14	100	8	8.00
	21	104	20	19.23
	28	112	22	19.64

Table 53: Mortality in *S. obliqua* Due to Viruses

Month	Date	No. of Larvae Collected	Viruses	
			Mortality	Per cent Mortality
August 1990	4	–	–	–
	11	–	–	–
	18	22	–	–
	25	24	3	12.5
September 1990	1	47	4	8.51
	8	51	7	13.72
	15	59	8	13.55
	28	63	9	14.28
	29	71	14	19.71
October 1990	6	97	20	20.61
	13	114	23	20.17
	20	112	3	2.67
	27	108	21	19.44

Table 54: Mortality Due to Grannulosis Virus Spray

Sl.No.	Replicates No. of Larvae Taken	S. obliqua		A. lactinea		T. postica	
		Mortality	Per cent Mortality	Mortality	Per cent Mortality	Mortality	Per cent Mortality
1.	150	18	12.00	12	8.00	48	32.00
2.	120	29	24.16	45	37.50	43	35.83
3.	100	37	37.00	20	20.00	25	25.00
4.	150	30	20.00	54	36.00	62	41.33
5.	150	84	56.00	74	49.53	89	59.33
6.	150	84	56.00	79	52.66	108	72.00
7.	200	120	60.00	118	59.00	137	68.50
8.	200	114	57.00	120	60.00	150	75.00
9.	150	62	41.33	38	25.53	110	73.33
10.	150	60	40.00	66	44.00	111	74.00

Table 55: Mortality in *A. lactinea* Due to Fungus

Month	Date	No. of Larvae Collected	Fungus Mortality	Fungus Per cent Mortality
August 1988	6	–	–	–
	13	–	–	–
	20	27	–	–
	27	26	–	–
September 1988	3	59	9	15.25
	10	63	11	17.46
	17	61	10	16.39
	24	78	13	16.86
October 1988	1	86	20	23.25
	8	92	24	26.08
	15	96	28	29.16
	22	101	32	31.63
	29	98	31	31.63

Table 56: Mortality in *A. lactinea* Due to Fungus

Month	Date	No. of Larvae Collected	Fungus	
			Mortality	Per cent Mortality
August 1989	5	4	–	–
	12	6	–	–
	19	18	1	5.55
	26	27	3	8.10
September 1989	2	41	5	12.19
	9	39	10	25.64
	16	53	12	22.64
	30	71	23	32.39
October 1989	7	89	21	23.54
	14	98	31	31.63
	21	104	30	28.84
	28	102	33	32.35

Table 57: Mortality in *A. lactinea* Due to Fungus

Month	Date	No. of Larvae Collected	Fungus Mortality	Fungus Per cent Mortality
August 1990	4	—	—	—
	11	—	—	—
	18	8	—	—
	25	16	1	6.25
September 1990	1	27	4	14.81
	8	29	5	17.24
	15	38	6	15.78
	22	49	15	30.61
	29	58	18	32.75
October 1990	6	69	11	15.94
	13	87	16	18.39
	20	102	29	28.43
	27	106	33	31.13

Table 58: Mortality in *A. lactinea* Due to NPV

Month	Date	No. of Larvae Collected	NPV Mortality	NPV Per cent Mortality
August 1988	6	–	–	–
	13	21	–	–
	20	38	5	13.15
	27	37	–	–
September 1988	3	39	7	17.94
	10	51	9	17.64
	17	69	19	27.53
	24	74	17	22.97
October 1988	1	62	14	22.58
	8	61	16	26.22
	15	66	18	27.27
	22	70	19	27.14
	29	71	20	28.16

Table 59: Mortality in *A. lactinea* Due to NPV

Month	Date	No. of Larvae Collected	NPV Mortality	NPV Per cent Mortality
August 1989	5	–	–	–
	12	8	1	12.5
	19	–	–	–
	26	19	3	15.78
September 1989	2	16	2	12.5
	9	31	3	9.67
	16	38	6	15.78
	23	38	6	15.78
	30	49	11	22.44
October 1989	7	51	10	19.60
	14	48	10	20.83
	21	69	19	27.53
	28	75	21	28.00

Table 60: Mortality in *A. lactinea* Due to NPV

Month	Date	No. of Larvae Collected	NPV Mortality	NPV Per cent Mortality
August 1990	4	–	–	–
	11	–	–	–
	18	–	–	–
	25	29	2	6.89
September 1990	1	31	3	9.67
	8	36	3	8.33
	15	39	4	10.25
	22	49	7	14.28
	29	56	9	16.07
October 1990	6	69	15	21.73
	13	62	14	22.58
	20	71	20	28.16
	27	74	18	24.32

T. postica

Granulosis virus was tested against 4th instar *T. postica* larvae. The results (Table 54) showed that Granulosis virus cause maximum 75 per cent mortality in the larvae and minimum 25 per cent.

Discussion

Bell (1981) studied the potential of use of microbials in management of *Heliothis* sp. He reported that all of the major groups of entomopathogens were with some potential for use in *Heliothis* management. The use of these microbials was very considerable between crops and locations, depending upon climate, disease symptomatology and economic thresholds of crop damage. In general, the microbials acted naturally in the environment as population suppressors, and as such, ideally as elements in integrated pest management (IPM) programmes. The majority of the research on microbial control of *Heliothis* have been conducted in cotton using the bacterium, *Bacillus thuringiensis* Berliner and nuclear polyhedrosis viruses. These pathogens are presently used to suppress low to moderate populations of larvae in cotton within the IPM framework. Recent studies have indicated that using gustatory-stimulant adjuvants can increase the effectiveness of the microbials and can result in the control of higher populations of *Heliothis* than normally feasible. However, the current use of such induced epizootics as single factor methods for control is negligible compared with the less costly chemical control methods. Although, the immediate future for sizable market of microbials apparently lies within the IPM programmes, the potential for use in unresearched areas remains high.

Ignoffo and Hostetter (1984) studied the microbial diseases in Cabbage looper, *Trichoplusiani* (Hubner). The symptoms and causative agent of disease of the cabbage looper have been reported. They isolated about 20 species of pathogens from the larvae of *T. ni*. Four viruses of three viral groups, *i.e.*, two nuclear polyhedrosis, one cytoplasmic polyhedrosis and one granulosis were isolated. Two bacteria (*B. thuringiensis, Serratia marscens* Bizio.) were isolated from laboratory reared larvae. Six protozoan species were reported from *T. ni*, two from naturally infected larvae (*Nosema trichoplusiae, Thelohania* sp). More fungi than any other group of pathogens have been isolated from *T. ni*. Those were two species of *Entomopthora*, two species of *Metarrhizium* and one each of *Nomuraea, Aspergillus, Beauveria,* and *Spicaria*. In the present study *Aspergillus* sp., *Entomopthora* sp. and *B. thuringiensis* were isolated from *S. obliqua* and *A. lactinea*.

A nuclear polyhedrosis virus was isolated from the red hairy caterpillar *Amsacta albistriga* (Walker) by I.I.H.R. Scientist (Anonymous, 1982). It is also noted that *S. obliqua* was attacked by *B. thuringiensis*. Battu *et al.* (1972) investigated the microbial infections in certain insect pest in Punjab. 21 isolates (15 Bacteria, 4 fungi, 2 viruses) proved pathogenic to their respective host in the laboratory. The host tried were *Mythimna separata* Walk., *Spodoptera litura* (Fab.) and *Plusia orichalcea*. Again, Battu *et al.* (1978) during 1974–76 surveyed the incidence of microbial infections of insect pests. Diseased insects found were chiefly lepidopterous larvae. Infections identified as granuloses viruses in *S. litura* and polyedrosis viruses in *S. litura, H. armigera* and *S. obliqua*. The incidence of these infections was sporadic and no epizootics were observed in the field.

Jacob and Thomas (1973) noted the mortality in the larvae of the polyphagus pest, *S. obliqua* after 6 to 8 days from the time of field collection date, due to Nuclear polyedrosis virus (NPV). The diseased larvae were collected from sweet potato fields at Vallayani, Kerala State, India in 1972. Jacob *et al.* (1973) also reported 2 virus diseases in *Pericallia ricini* Fabr. (Arctiidae : Lepidoptera) *viz.* a nuclear polyhedrosis and a granulosis virus. Infected larvae were found infesting castor plants (*Ricinus communis*) at Vallayani, Kerala State, India in July 1972. Mixed infections occurred and death from the two viruses took place in 6–10 and 8–12 days respectively. The symptoms were also described.

Laboratory studies were carried out by Kamat *et al.* (1978) on the feasibility of the use of the fungus, *Nomuraea rileyi* for the biological control of *Achaea janata* (L.) an important pest of castor.

The fungus was isolated from dead larvae and a suspension and dust were prepared from the culture. Larvae in petridishes or on healthy castor plants were sprayed with the spore suspension and examined daily. They became sluggish within a week of treatment, inactive 12–14 days after it, and by the end of 2 weeks 80–90 per cent had died. Two applications were more effective than 1 and the dust formulation was more effective than the suspension causing upto 100 per cent mortality. The fungus was not pathogenic for the parasite, *Telenomus proditor* Nixon which attacked the eggs of *A. janata*.

Field studies were carried out by Krishnaiah *et al.* (1981) in Karnataka between 1974–1978 to evaluate the effectiveness of *B. thuringiensis* var *Kurstaki* (Dipel) and another formulation of *B. thuringiensis* (Cajral) against various pests including *Adisura atkinsoni* Moore on trailing bean (*Lablab purpureus*) (*Peliches Lablab*). Dipel was ineffective against *A. atkinsoni*. Mathur and Mathur (1968) have made observations on parasitic *Aspergillus* on larvae of *A. moorie* in which they found that late stage larvae of *A. moorie* (Btl), an important pest of crops in Rajasthan, India were found infected with *Aspergillus funigatus* in the field in September 1966. In laboratory tests, no larvae were infected before the fourth instar but, the fungus was highly pathogenic to older ones.

Mistry *et al.* (1985) evaluated the nuclear polyhedrosis virus of the noctuid, *H. armigera* against the pests infesting tomato and chickpea (*Cicer arietinum*) in Gujarat. India, in 1980–82. Five spray applications of the virus at 250 larvae equivalants per hectare per week gave satisfactory control of the pest and resulted in a grain yield increase of 28 per cent in chickpea in 1980–81 and 47 per cent in 1981–82. Nair and Jacob (1976) concluded that light as temperature is responsible for the inactivation of the virus in the field. The virus caused no symptoms in larvae of 4 other species of lepidoptera against which it was tested. In the present study, at laboratory experiments highest mortality due to granulosis virus spray was 60 per cent in *S. obliqua* and *A. lactinea* while it was 75 per cent in *T. postica*.

Nene (1973) reported that the fungus *Paecilomyces farinosus* was found infecting adults of *Bemisia tabaci* (Gennadius) on cotton in the field in U.P. in 1971. Mortality was high and the incidence of yellow mosaic virus disease on soyabean, green gram, black gram and okra on which the Aleyrodid was a vector, was less than 10 per cent as compared with 100 per cent in some varieties in previous years. In the laboratory the fungus caused 90 per cent mortality in adults of *B. tabaci.*

Patel *et al.* (1978) observed and recorded milky disease in white grubs (*Holotrichia* sp) in India. Diseased larvae collected from Kapadvanj in the autumn of the year 1968 were found to be infected with a local milky disease. This was the first record of such a disease from India and the casual organism appeared to be different from others causing similar disease elsewhere. In subsequent tests in 1969, healthy larvae were successfully artificially infected with the disease.

Rabindra and Subramanian (1974) reported an epizootic of a nuclear polyhedrosis virus in *Dasychira mendosa* (Hb.) on castor was reported from Coimbatore, in India, in 1973. It was the first record of a NPV for this species although, similar diseases were known in other species of *Dasychira.* The polyhedra dimensions of which are given were found in the nuclei of the fat body hypodermis and tracheal matrix of *D. mendosa.* Rabindra *et al.* (1975) noted that NPV was found in larvae of *Chrysodeixis chalcites* (Esp) (*Plusia chalcytes*) feeding on groundnut and *Flaveria australasia* at Coimbatore. It had an incubation period of 5–7 days in fourth instar larvae and caused rupturing of the integument and hypertrophy of the nuclei of the midgut epithelial cells. It was the first record of NPV in *C. chalcites* from India, although a similar virus was reported in this species from France.

Ramakrishnan *et al.* (1975) reported the virions and inclusion bodies of a nuclear polyhedrosis virus affecting *S. litura* in India. This was the first true record of a polyhedrosis virus from this species in India. Since a previous report of such a virus was concerned with *S. littoralis* (Boised) and not *S. litura*.

Ramaseshiah (1973) observed the fungus *Entomopthora grylli* Fres attacking and killing larvae of *S. obliqua* on groundnuts at Bangalore in November 1972. Climatic conditions in that month favoured the development of an epizootic and large numbers of larvae were killed. He also pointed out few records of this fungus attacking Arctiids. *E. grylli* was also isolated from a dead larvae that could not be identified, collected on *Calotropis gigantea* at Hessaraghatta in December 1972.

Kulshreshtha *et al.* (1965) tested 10 per cent BHC dust and dust containing 10^8 spores of *B. thuringiensis* per gram, of both gave 88–89 per cent mortality in 96 hrs of larvae of *A. janata* in the third, fourth and fifth instars, but the BHC dust was much less effective against the sixth instar, whereas, the *Bacillus* gave nearly 85 per cent kill. In a field test on castor the BHC dust gave 52.3 per cent kill in 24 hrs and the *Bacillus* 73.91 per cent.

During a survey in India for fungi pathogenic to pests of sugar crops, *S. obliqua* was found to be infested by a species of fungus *Conidiobolus thromboides* (*Entomopthora virulenta*) at Lucknow, U.P. (Verma *et al.*, 1982). Symptoms of the disease were described. Some diseased larvae were observed to have their normal food plant and climb to the top of grasses or pigeonpea plants. They observed incidence of infection was 94 per cent in the fields and 100 per cent in the laboratory and it appeared to be favoured by heavy rain, frequent irrigation or high population density.

The larval susceptibility to nuclear polyhedrosis virus (NPV) and granulosis virus (GV) in the high and low density forms of *Pseudaletia separata* was examined under laboratory condition by Kumini and Yamada (1988). With the rearing density decreased, the larvae were more susceptible and died earlier when 5th instar larvae were perorally inoculated with NPV. The larvae reared in isolation were more susceptible than those reared in crowd when 2nd instar larvae were perorally inoculated with GV. The pale larvae produced from crowded rearing were more susceptible than the black larvae when NPV was inoculated perorally. No differences in susceptibility and duration from inoculation to death were observed between the isolated larvae and the crowded larvae when NPV was inoculated subcutaneously.

The variability of 10 isolates each of *Beauveria bassiana* and *Metarrhizium anisopliae* was investigated by Searle and Yule (1988) for response to temperature, relative humidity and pH. Differences in response were found within, as well as between species. Each isolate was tested for pathogenicity on each soil inhabiting stage of the carrot weevil under optimal and sub-optimal laboratory conditions. LD and LT_{50} values varied significantly between the most and least virulent isolates of each species.

INTRASPECIFIC RELATIONSHIP (CANNABALISM)

Introduction

Competition is considered to apply only where the organisms search for and required some common resource or other real thing that is in short supply. DeBach and Sundby (1963) says that competition may occur when no such resource is in short supply. The competitive displacement can result even when no real resource is competed for (Huffaker and Messenger, 1964).

Application of the competitive displacement principle has resulted in many improvements in pest control, especially in weed control. There are also many examples of displacement of previously existing natural enemy by one later introduced, in every case resulting in a lowered density of the pest. Intraspecific and interspecific competition are quite distinct, the former is inherently a regulating process. Decreasing density of a species 'A' commonly means a decreased chance of its recovery as a population so far as its competition with species 'B' is concerned. A decrease in numbers when all the competition is intraspecific, however, means automatically a lessening of the effects of that competition and improved chances of recovery.

Intraspecific competition occurs due to the want of food, shelter and in search of opposite sex, therefore, insects can controlled by depriving them of their food, shelter and mates. The intraspecific response promotes population growth very restraining. Cannibalism or intraspecific predation. is no longer regarded as a deviation from normal behaviour in natural populations. Cannibalism is not confined to laboratory populations nor to otherwise abnormally stressed populations. There are obvious potential benefits of cannibalism such as obtaining an additional food source and eliminating a possible competitor or predators. In past, such competitions with respect to their population dynamics of insects have been studied by Mathur (1966), Ischaque *et al.* (1985), Sathe *et al.* (1988) and others.

Materials and Methods

Laboratory cultures of *S. obliqua, A. lactinea* and *T. postica* were used for the experiments dealing with cannibalism. 200 caterpillars were kept in each glass throughs for 12 hr. Experiments were conducted with control and by providing abundant food. The observations were made on 200 caterpillars in each experiment and the experiments were replicated for 5 times. One to five instars of each species of the pest were considered for the studies. In control, the individuals provided no food.

Results

S. obliqua

Cannibalism was not observed in 1st and 2nd instars of *S. obliqua* larvae. Cannibalism reached in its peak *i.e.* 20 per cent in 5th instar, and in 3rd, 4th instars, the percentage mortality due to cannibalism was 7 per cent, 17.5 per cent respectively. The cannibalism rate was quite low when the individuals provided abundant food. Results are tabulated in Tables 61 and 62.

Table 61: Cannibalism in *S. obliqua*

Instars	Total No. Tried	Total No. of Individuals Died	Per cent Mortality
1st Instar	200	00.00	00.00
2nd Instar	200	00.00	00.00
3rd Instar	200	14.00	7.00
4th Instar	200	35.00	17.5
5th Instar	200	40.00	20.00

A. lactinea

The highest mortality due to cannibalism in larvae of *A. lactinea* was 18 per cent in 5th instars when the larvae were starved. In 3rd and 4th instars mortality was 7.5 per cent and 16.5 per cent respectively. In first two instars cannibalism was not observed. Results are recorded in Tables 63 and 64.

Table 62: Control Cannibalism in *S. obliqua* with Food

Instars	Total No. Tried	Total No. of Individuals Died	Per cent Mortality
1st Instar	200	00.00	00.00
2nd Instar	200	00.00	00.00
3rd Instar	200	4.00	2.00
4th Instar	200	9.00	4.50
5th Instar	200	11.00	5.50

Table 63: Cannibalism in *A. lactinea*

Instars	Total No. Tried	Total No. of Individuals Died	Per cent Mortality
1st Instar	200	00.00	00.00
2nd Instar	200	00.00	00.00
3rd Instar	200	15.00	7.50
4th Instar	200	33.00	16.50
5th Instar	200	36.00	18.00

Table 64: Control Cannibalism in *A. lactinea* with Food

Instars	Total No. Tried	Total No. of Individuals Died	Per cent Mortality
1st Instar	200	00.00	00.00
2nd Instar	200	00.00	00.00
3rd Instar	200	2.00	1.00
4th Instar	200	5.00	2.50
5th Instar	200	5.00	3.50

Discussion

According to Fox (1975) and Polis (1981) cannibalism is a natural feature of numerous species and that it can have profound consequences on population dynamics. Cannibalism has been reported in the corn earworm, *Heliothis zea* (Boddie). Since the last century (Mally, 1892) *H. zea* is a widespread species of economic importance. Corn, cotton, soyabean, tomato and tobacco among its many host plants. Corn is one of the preferred host of *H. zea*, fresh silks were preferred oviposition site. Larvae move from the silks to the tip of the ear where they feed on the developing earhead. Mally (1892) noted that cannibalism in *H. zea* can reduce the size of a population. He observed that many eggs were laid on the silks of an ear of corn. Cannibalism often resulted in only one larval development in one corn ear. Stinner *et al.* (1976) reported that it was not unusual for mortality due to cannibalism to exceed 75 per cent.

Barber (1936), Hamilton (1970) and Stinner *et al.* (1976) reported in ears of corn that cannibalism occur even though there was sufficient food to support more than one larvae to maturity. To explain this high mortality in the absence of food resource limitation Hamilton (1970) and Stinner *et al.* (1976) theorized that cannibalism may have evolved when the predominant hosts of *H. zea* were with flower heads or small ancestral corn ears inadequate for the successful development of more than one larvae. Stinner *et al.* (1976) suggested that cannibalism in corn feeding *H. zea* was an evolutionary relict of no benefit. In other species, Fox (1975) and Polis (1981) reported that cannibalism yields nutritional benefits, especially when food was scarce. These benefits were typically manifested by accelerated development or increased productivity. Cannibalism can also profoundly affect population dynamics (Fox, 1975; Polls, 1981). Stinner *et al.* (1976) described some important effects of cannibalism on the timing of population cycles and the density of *H. zea.* They reported that cannibalism limits the population size.

Mathur (1966) reported the cannibalism in *A. moorei* (Lepidoptera) while conducting ecological studies. The extent of cannibalism in the species has been experimentally assessed by him. He found that the caterpillars feed on each other inspite of the presence of sufficient food. Before the larvae are induced to feed on each other the flickering and warding off reactions took place which were very common at the time of moulting. Generally, larvae started feeding from the posterior region of the abdomen. The 1st and 2nd instars were completely devoured by the 4th and 5th instar larvae. He considered 100 insects of each instar in four replicates. Further, he reported that the cannibalism did not occur among the 1st instar larvae and it was negligible in 6th instar. The extent of cannibalism among the insects of 2nd and 5th instars separately was almost the same ranging between 5 to 6 per cent. It was highest among the 4th instar (16 to 20 per cent) followed by 3rd instar (12 to 14 per cent). In the mixed culture of the 4th and 2nd instar larvae, the 4th instar larvae devoured 80 to 90 per cent larvae of the 2nd instar. In the present study negligible cannibalism was recorded when provided abundant food to the caterpillars. But, it reached 20 per cent, 18 per cent in *S. obliqua* and *A. lactinea* respectively when provided with specific density of preys.

Joyer and Gould (1985) studied the developmental consequences of cannibalism in *H. zea* (Lepidoptera). They reported that on a standard artificial diet, there were no significant differences between cannibals and non-cannibals. These results indicate a nutritional benefit of cannibalism under less than optimal (low moisture) conditions that was not found under better conditions. However, they concluded that the countless environmental interactions may dictate the degree to which cannibalism was beneficial or detrimental to individuals and populations of *H. zea*. At the same time that cannibalism increased mortality in a population, it may also result in robust individuals more resistant to other causes of mortality and more fecund.

The cannibalistic activity among the adults of three different species of grasshoppers was studied separately at different conditions of temperature, food and density by Ishaque *et al.* (1985). They reported that among the fully fed adults the process was found accelerated by higher temperature and density and the highest rate of cannibalism was obtained in *Oedaleus abruptus* followed by *Spathosternum prasinferum* and *Hieroglyphus nigrorepletus*. Similarly, among the starved adults, the process was accelerated by higher temperature, density along with the duration of starvation and highest rate was recorded in *H. nigrorepletus* followed by *O. abruptus* and *S. prasiniferum*. The rate of cannibalism was found higher under starvation than under fully fed conditions. Similar observations were noted in the present forms studied.

Recently, Sathe *et al.* (1988) worked cannibalism in *H. armegira* wherein they noted that the caterpillars feed on each other inspite of the presence of sufficient food. Before the larvae induced to fed on each other the flickering and warding off reaction took place which were very common at the time of moulting. The 1st and 2nd instars were completely devoured by the 4th and 5th instar larvae. The cannibalism did not occur among the 1st instar larvae and it was negligible in 6th instar probably due to the closer pupal period wherein they might stop feeding. The percentage of cannibalism among the instars of 2nd, 3rd, 4th and 5th were 2, 8, 23, 17 per cent respectively. Control cannibalism was much higher than with food. Maximum 76 to 82 per cent and 23 per cent cannibalisms was noted with control and food respectively on 4th instar. In the mixed culture of the 4th and 2nd instar larvae, the 4th instars devoured about 94 per cent larvae of the 2nd instar. In the present forms the cannibalism rate was not so high (20 per cent in *S. obliqua* and 18 per cent in *A. lactinea*) as compared to *H. armigera*.

Bibliography

Note: *W.L. indicates the number of the periodical in the "World List of Scientific Periodicals" 4th Edn. (1963–65) edited by P. Brown and C.B. Stration. This number is given at the first citation of the periodical only.*

Agarwala, B.K., Datta, S. and Raychaudhauri, D.N., 1983. An account of Syrphid (Diptera : Syrphidae) predators of aphids available in Darjeeling District of West Bengal. *Pranikee*, 4: 238–244.

Ali, S.A.E., 1961. Campaign against red hairy caterpillar in Madras State in 1958 and 1959. *Pl. Prot. Bull.*, New Delhi, 13: 38–40.

Anand, M. and Pant, J.C., 1986. Influence of vitamins of B-complex on the growth and survival of *Chilo partellus* (Swinhoe) larvae. *Indian J. Ent.*, 48: 241–243 (W.L. 22997).

Anonymous, 1953. An unusual epidemic of the castor hairy caterpillar, *Euproctis lunata* Walker, in Delhi. *Pl. Prot. Bull.*, New Delhi, 5: 50–51.

Anonymous, 1982. Annual Report: All India Coordinated Research Project on *Meterological Control of Crop Pests and Weeds*. April 1981 to March 1982, pp. 1 (IIHR) 24 (GAU).

Anonymous, 1985. Annual Report: All India Coordinated Research Project on *Biological Control of Crop Pests and Weeds*. January, 1984 to December 1984, pp. 71–82.

Anonymous, 1986. Annual Report: All India Coordinated Research Project on *Biological Control of Crop Pests and Weeds*. January, 1985 to December, 1985, pp. 63–68.

Arthur, A.P., 1963. Life histories and immature stages of four Icheumonid parasites of the European pine shoot moth, *Bhyacionia budiana* (Schiff) in Ontario. *Can. Ent.*, 95: 1078–1011 (W.L. 13141).

Ayyar, P.N.K. and Narayanaswami, P.S., 1940. On the biology of *Spathius vumeficus* Wilk, a possible effective parasite of *Pempheres affinis* in South India. *Indian J. Ent.*, 2: 79–86.

Bains, S.B. and Shukla, K.K., 1976. Effect of temperatures on the development and survival of maize borer. *Chilo partellus* (Swinhoe). *Indian J. Ecol.*, 3: 149–155.

Bakhetia, D.R.C. and Sidhu, A.S., 1977. Biology and seasonal activity of the groundnut aphid, *Aphis craccivora* Koch. *J. Res. PAU*, 14: 299–303.

Bank, C.J. and Macaulay, E.D.M., 1970. Effects of varying the host plant and environmental conditions on the feeding and reproduction of *Aphis fabae* Scop. *Entomologia Exp. Appli.*, 13: 85–86 (W.L. 18217).

Barber, G.W., 1936. The Cannibalistic Habits of the Corn Earworm. *U.S. Dep. Agri. Tech. Bull.*, 49: 1–11.

Battu, G.S., Bindra, O.S. and Rangarajan. M., 1972. Investigations on microbial infections of insect pests in the Punjab. *Indian J. Ent.*, 33: 317–325.

Battu, G.S. and Dhaliwal, G.S., 1976. *Carcalia* sp. (Tachinidae : Diptera) as a parasite of *Diacrisia obliqua* (Walker) in Punjab. *Current Research*, 5: 122–123.

Battu, G.S., Dilawari, V.K. and Bindra, O.S., 1978. Investigations on the microbial infections of insect pests in the Punjab II. *Indian J. Ent.*, 39: 271–280.

Bell, M.R., 1981. The potential use of microbials in *Heliothis* management. In: *Proc. International Workshop on Heliothis Management*, pp. 137–145.

Bell, M.R. and Romine, C.L., 1980. Tobacco budworm field evaluation of microbial control in cotton using *Bacillus thuringiensis* and a nuclear polyhedrosis virus with a feeding adjuvant. *J. Eco. Entomol.*, 73: 427–430. (W.L. 25921).

Berisford, C.W., Kulman, H.M. and Pienkowshi, R.L., 1970. Notes on the biologies of hymenopterous parasites of IPS spp. bark beetles in Virginia. *Can. Ent.*, 102: 484–490.

Bhalani. P.A., 1989. Suitability of host plants for growth and development of leaf eating caterpillars *Spodoptera litura* (Fabr.). *Indian J. Ent.*, 51: 427–430.

Bhatia, P. and Sethi, G.R., 1989. Studies on the nutritional variables in *Tribolium castaneum* (Herbst) reared on different varieties of Bengal gram, *Cicer arietinum* Linn. *Indian J. Ent.*, 51: 29–38.

Bhatnagar, S.P., 1948. Studies on *Apanteles* Foerster (Vipionidae : Parasitic hymenoptera) from India. *Indian J. Ent.*, 10: 133–210.

Bhatnagar, N.S., 1977. Record of the parasites and predators of Soyabean leafrollers in India. *Indian J. Ent.*, 38: 383–384.

Bilapate, C.G., Raodeo, A.K. and Pawar, V.M., 1979. Population dynamics of *Heliothis armigera* (Hubner) on sorghum, pigeonpea and chick pea in Marathwada. *Indian J. Agric. Sci.*, 49: 560–566.

Bilapate, G.G. and Pawar, V.M., 1980. Life fecundity tables for *Heliothis armigera* Hubner (Lepidoptera : Noctuidae) on Sorghum earhead. *Proc. Indian Acad. Sci. (Ani. Sci.)*, 89: 69–73 (W.L. 39255).

Bindra, O.S. and Kittur, S.U., 1961. Biology and control of the red hairy caterpillar (*Amsacta moorei* Butler) in M.P. *Vikram Sci.*, 5: 62–71.

Birch, L.C., 1948. The intrinsic rate of natural increase in an insect population. *J. Anim. Ecol.*, 17: 15–26 (W.L. 25559).

Bose, K.C. and Ray, S.K., 1967. Aphid predators balance. II. Comparative study on the consumption of aphids by the common predator, *Chilomanes saxmaculata* Fabr. (Coleoptera) (Coccinellidae). *Indian J. Sci., Industr.*, 1: 56–59.

Broodryk, S.W., 1969a. The biology of *Chelonus* (*Microchelonus*) *curvimaculatus* Cameron (Hymenoptera : Braconidae). *J. Ent. Soc., South Afirca*, 169–189 (W.L. 25972).

Broodryk, S.W., 1969b. The biology of *Orgilus parcus* Turner (Hymenoptera : Braconidae). *Ibid*, 32: 243–257.

Butani, P.G. and Bharodia, R.K., 1984. Relation of a ground nut aphid population with its natural predator, lady bird beetles. *Gujarat Agri. Univ. Res.*, 9: 72–74.

Butter, G.D. Jr. and Scott, D.R., 1976. Two models for development of the corn earworm on sweet corn in Idaho. *Environ. Entomol.*, 5: 68–72.

Cals, P. and Shaumer, N., 1965. Bilogie et morphologic larvaire comparees de trois Ichneumonidae pimplines. *Annl. Sci. Nat. Zool.*, 7: 767–790.

Calvert, D.J. and R. Vanden Bosch, 1972. Behaviour and biology on *Monoctonus paulensis* (Hymenoptera : Braconidae), a parasite of dactynotine aphids. *Ann. Ent. Soc. Am.*, 65: 773–779 (W.L. 3145).

Cardona, C. and Oatman, E.R., 1971. Biology of *Apanteles dignus* (Hymenoptera : Braconidae), a primary parasite of the Tomato pin worm. *Ann. Ent. Soc., Am.*, 64: 996–1007.

Chakravarthy, A.K. and Lingappa, S., 1979. Wagtails as predators of field bean aphids. *J. Bombay Nat. Hist. Soc.*, 76: 367 (W.L. 25676).

Chand, P., 1979. Polyhedrosis of *Diacrisia obliqua* Walker. *Indian J. Ent.*, 41: 194.

Chand, P. and Prasad, D., 1970. A new natural enemy of *Diacrisia obliqua* Walker. *Indian J. Ent.*, 40: 359.

Chandra, H., Venkatesh, M.V. and Ahluwalla, P.T.S., 1985. Effect of some food plants on the maturation, fecundity and fertility of *Schistocerca gregaria* Froskal. *Indian J. Ent.*, 47: 354–355.

Chandra, H. and Kushwaha, K.S., 1987. Impact of environmental resistance on aphid complex of cruciferous crops under the agroclimatic conditions of Udaipur. II. Biotic components. *Indian J. Ent.*, 49: 86–113.

Chapman, R.N., 1925. *Animal Ecology.*

Cherian, C. and Narayanswami, 1942. The biology of *Microbracon chilonis* Viereck, a larval parasite of *Chilo zonellus* (Swin). *Indian J. Ent.*, 4: 1–4.

Chhibber, R.C., 1975. Know the insect pest of sugarbeet. *Indian Farmer's Digest*, 8: 21–24.

Chumakova, B.M., 1959. Entomophages of San Jose scale in USSR and ways to increase their effectiveness. In: *1st Int. Conf. Insect. Pathol. and Biol. Control,* Prague (in Russian), pp. 481–485.

Clausen, C.P., 1940. *Entomophagous Insects.* McGraw Hill, New York, pp. 688.

Cole, L.R., 1970. Observations on the finding of mates by male *Phaeogenes invisor* and *Apanteles medicaginis. Anim. Behav.*, 18: 184–189.

Coppel, H.C. and Mertins, J.W., 1977. *Biological Insect Pest Suppression.* Springer-Verlag, Berlin Heidelberg. New York, pp. 1–314.

C.S.I.R. Authors 1980. *Agricultural Entomology.*

Dabi, R.K., Mehrotra, P. and Shinde, V.K.R., 1980. Bioefficacy of different levels of *Bacillus thuringiensis* Berliner against *Diacrisia obliqua* Walker. *J. Ent. Res.*, 4: 231–233.

Dakshayani, K., Bentur, J.S. and Kalode, M.B., 1988. A meridic diet for rice leaf folder. *Entomon.*, 13: 309–312 (W.L. 18270).

David, B.V. and Basheer, M., 1961. Mass occurrence of the predatory stink bug, *Cantheconidia* (Canthecona) *furcellata* (Wolff) in South India. *J. Bombay Nat. Hist. Soc.*, 58: 817–819.

DeBach, P. and Sundby, R.A., 1963. Competitive displacement between ecological homologues. *Hilgardia*, 34: 105–166 (W.L. 22186).

Deevey, E.S. Jr., 1947. Life tables for natural populations of animals Quart. *Rev. Bid.*, 22: 283–314.

Deshmukh, P.D., Rathore, Y.S. and Bhattacharya, A.K., 1977. Studies on the growth and development of *D. obliqua* (Walker) (Lep. Arctiidae) on sixteen plant species. *Z. Ang. Ent.*, 84: 431–435.

Deveraj, K.C. and Subramanya, R.V., 1987. Effect of antibiotics on growth, development and life cycle of *Spodoptera litura* Fab. (Lepidoptera : Noctuidae). In: *Proc. III Orient. Ent. Sym.*, pp. 83–88.

Dhiman, S.C., 1986. Effect of temperature on the seasonal occurrence of *Cletus signatus* Walker (Heteroptera : Coreidae). *Proc. I Orient. Ent. Symp.*, pp. 91–94.

Dowden, P.B., 1934. *Zenillia lebatric* Panzer, a tachind parasite of the gypsy moth and brown-tail moth. *J. Agric. Res.*, 48: 97–114.

Dowell, R.V. and Horn, D.J., 1975. Mating behaviour of *Bathypletes curculionis* (Hymenoptera : Ichneumonidae) a parasitoid of the alfalfa weevil. *Hypera postica* Col. (Curculionidae). *Entomophaga*, 20: 271–273 (W.L. 18271).

Ellington, J.J., 1970. Approaches to a more meaningful evaluation of host plant resistance to lepidoptera. *Beltwide Cotton Prod. Res. Conf. Rpt.*, pp. 80–81.

Eubank, W.P., Atmar, J.W. and Ellington, J.J., 1973. The significance and thermodynamics of fluctuating varsus static thermal environments on *Heliothis zea* egg development rates. *Environ. Entomol.*, 772: 491–496.

Falcon, L.A., 1974. Insect pathogens, integration into a pest management system. In: *Proceedings, Summer Institute on Biological Control of Plant Insects and Diseases*. Miss Jackson. University Press of Mississippi, USA, pp. 618–627.

Faleiro, J.R., Singh, K.M. and Singh, R.N., 1986. Pest complex and succession of insect pest in cowpea, *Vigna unguiculata* (L.) Walp. *Indian J. Ent.*, 48: 54–61.

Fisher, R.C., 1959. Life history and ecology of *Horogenes chrysostictes* Gmelin (Hymenoptera : Ichneumonidae) a parasite of *Ephestia sericurium* Scott. (Lepidoptera : Phycitidaes). *Can. J. Zool.*, 37: 429–446 (W.L. 13191).

Fox, L.R., 1975. Cannibalism in natural population. *Ann. Rev. Ecol. Syst.*, 6: 87–106.

Fulton, B.B., 1940. The hornworm parasite, *Apanteles congregatus* (Say) and the hyperparasite, *Hypopteromalus tabacum* (Fitch). *Ann. Ent. Soc. Am.*, 33: 231–244.

Fye, R.E. and Poole, H.K., 1971. The effect of high temperatures on the fecundity and fertility of six lepidopterous pests of cotton in Arizona. *U.S. Dept. Agr. Prod. Res. Rpt.,* 131: 8.

Fye, R.E. and McAda, W.C., 1972. Laboratory studies of the development, longevity and fecundity of six species of lepidopterous pests of cotton in Arizona. *U.S. Dept. Agr. Tech. Bul.,* 1454: 71.

Fye, R.E. and May, C.J., 1974. Development, fecundity and longevity of the cotton leaf perforator in relation to temperature. *U.S. Dept. ARS.,* W12: 1–5.

Ghosh, D. and Chakrabarti, S., 1986. Predatory complex of major aphids in the plants of West Bengal. *Proc. III Orient. Ent. Symp.,* pp. 177–182.

Goel, S.C. and Kumar, A., 1989. Seasonal build up of insect pests on a monsoon crop of sunflower in an agroecosystem. *Indian J. Ent.,* 53: 458–464.

Gordh, G. and Hendrickson, R., 1976. Courtship behaviour in *Bathyplectes anurus* (Thompson) (Hymenoptera : Ichneumonidae). *Ent. News,* 87: 271–279 (W.L. 18227).

Goyal, S.P. and Rathore, V.S., 1988. Patterns of insect-plant relationship determining susceptibility of different hosts to *Heliothis armigera* Hubner. *Indian J. Ent.,* 50: 193–201.

Grewal, S.S. and Atwal, A.S., 1969. The influence of temperature, humidity and food on the development and survival of *Earias insulana. J. Res. Punjab Agri. Univ.,* 6: 245–254.

Gupta, R.L., Pandey, N.P., Ram, S. and Sukhani, T.R., 1966. Bionomics of *Amsacta moorei* Butler (Lepidoptera : Arctiidae). *Allahabad Fmr.,* 40: 65–70.

Gupta, V.K., 1976. Host selection in Indian Ichneumonidae in relation to the ecology of the host. *Proc. Symp. Problem of Host Specificity in Insects,* pp. 91–96.

Gupta, V.K., 1988. Parasitic hymenoptera research and education during the 1980's. In: *Advances in Parasitic Hymenoptera Research,* pp. 1–7.

Hamilton, W.D., 1970, Selfish and spiteful behaviour in an evolutionary model. *Nature [London],* 228: 1218–1220 (W.L. 34258).

Hopper, K.R. and King, E.G., 1984. Preference of *Microplitis croeipes* (Hymenoptera : Braconidae) for instars and species for *Heliothis* (Lepidoptera: Noctuidae). *Environ. Entomol.,* 15: 1145–1150.

Howe, R.W., 1952. The rapid determination of intrinsic rate of increase of an insect population. *Appl. Biol.,* 40: 134–155.

Howe, R.W., 1967. Temperature effect on embryonic development of insects. *Ann. Rev. Ent.,* 12: 15–42 (W.L. 3436).

Huffaker, C.B. and Messenger, P.S., 1964. The concept and significance of natural control. In: *Biological Control of Insect Pests and Weeds,* (Ed.) P. DeBach. Reinhold, New York, p. 74–117.

Ignoffo, C.M. and Hostetter, D.L., 1984. Suppression and management of cabbage looper populations diseases. *Tech. Bull.,* 1684: 45–55.

Inamdar, S.A., 1991. Biosystemic studies in Braconid Parasitoids of some economic important crop pests from Western Maharashtra. *Ph.D. Thesis,* Shivaji University, Kolhapur, pp. 1–216.

Ishaque, M.B., Qamar, M. and Aziz, S.A., 1985. Cannibalism in *Spathosternum prasiniferum* Walker. *Oedaleus abruptus*, Thunberg and *Hieroglyphus nigrorepletus* Boliver under varying ecological conditions. *Indian J. Ent.*, 47: 437–443.

Jacob, A. and Thomas, M.J., 1973. A nucleus polyhedrosis virus of *Diacrisia obliqua* (Wlk). (Arctiidae : Lepidoptera). *Agri. Res. J. Kerala*, 10: 182.

Jacob, A., Thomas, M.J. and Chandrika, S., 1973. Occurrence of two virus diseases in *Pericallia ricini* Fabr. (Arctiidae : Lepidoptera). *Agri. Res. J., Kerala*, 10: 65.

Jagadish, P.S. and Channabasavanna, G.P., 1986. Effect of temperature and humidity on development of *Typhlodromps tetranychivorus* Gupta (Acari : Phytoseidae). *Proc. III Orient. Ent. Symp.*, 1: 169–172.

Jalali, S.K., Singh, S.P. and Chandis, R.B., 1987. Studies of host age preference and biology of exotic parasite, *Cotesia marginiventris* (Cresson) (Hymenoptera : Braconidae). *Entomon.*, 12: 59–62.

Jayanthi, R. and Metha, U.K., 1987. Performance of sugarcane varieties in relation to internode borer. *Chilo sacchariphagus* (K.) infestation. *Proc. III Orient. Ent. Sym.*, p. 73–78.

Jotwani, M.G., Sarup, P. and Pradhan, S., 1961. Effect of some important insecticides on the predator, *Stethorus pauparculus* Weise (Coccinellidae : Coleoptera). *Indian J. Ent.*, 22: 272–276.

Joyner, K. and Gould, F., 1985. Development consequences of cannibalism in *Heliothis zea* (Lepidoptera : Noctuidae). *Ann. Ent. Soc. Am.*, 78: 24–28.

Kadu, N.R., Radke, S.G. and Borle, M.N., 1987. Effect of temperatures on the development of *Heliothis armigera* (Hubner). *Indian J. Ent.*, 49: 535–543.

Kajita, H. and Drake, E.F., 1969. Biology of *Apanteles chilonis* and *Apanteles flavipes* parasites of *Chilo suppressalis*. *Mushi*, 42: 163–174 (W.L. 33872).

Kamath, S.M., 1960. Studies on the feeding habits, development and reproduction of the predaceous mite, *Phytoseiulus persimilis* (Acarina : Phytoseiidae) on some *phytophagus* mites in India. *Tech. Bull. Comm. Inst. Biol. Cont.*, 10: 49–56.

Kamat, M.N., Bagal, S.R., Thobbi, V.V., Rao, V.G. and Phadke, C.H., 1970. Biological control of castor semilooper through the use of entomogenous fungus, *Namuraea rileyi*. *Indian J. Bot.*, 1: 69–74.

Kapoor, K.N., Gujarati, J.P. and Gangrade, G.A., 1975. *Canthoconidia furcellata* Wolff. as a predator of *Prodenia litura* Fabr. larvae. *Indian J. Ent.*, 35: 275.

Katiyar, O.P., Lal, L. and Mukherji, S.P., 1975. Response of newly hatched caterpillars of *Diacrisia obliqua* to certain host plants. *Indian J. Ent.*, 37: 57–59.

Kaushik, S.K. and Naresh, J.S., 1989. Sampling for estimation of larval population of *Heliothis armigera* (Hubner) on chickpea. *Indian J. Ent.*, 51: 39–44.

Khan, M.Q. and Rao, A.S., 1948. Annual report of the scheme for research on the pests and diseases of castor and other oilseeds in Hyderabad State (1943–45), 67 pp. Dep. Agric, H.E.H. Nizam's Government. Hyderabad Dvn, Government Press.

Khan, R.M. and Sharma, S.K., 1972. *Cataglypis bicolor* Fab. (Hym. Formicidae) as a predator on few noctuid larvae. *Mds. Agri, J.*, 59: 192.

Khan, M.Z. and Hajeta, K.P., 1987. Studies on *Aulacophora foveicollis* Lucas (Coleoptera : Chrysomelidae) Food preference and extent of damage. *Indian J. Ent.*, 49: 457–459

Krishnaiah, K., Mohan, N.J. and Prasad, V.G., 1981. Efficacy of *Bacillus thuringiensis* for the control of lepidopterous pests of vegetable crops. *Entomon.*, 6: 87–93.

Kulshrestha, J.P., Sanghi, P.K. and Ravindranath, V., 1965. Microbial control of castor semilooper, *Achaea janata*. *Indian J. Ent.*, 27: 353–354.

Kulshrestha, J.P. and Agarwal, A.K., 1982. Predators and parasites-enemies of crop pests and farmers friend. *Pesticides*, 16: 3–6.

Kumar, N.S. and Ananthakrishnan, T.N., 1984. Predator-thrips interactions with reference to *Orius maxidentex* Chauri and *Carayonocoris indicus* Muraleedharan (Anthocoridoe : Heteroptera). In: *Proc. Ind. Nat. Sci. Acad.*, B50: 139–145.

Kunimi, Y. and Yamada, E., 1988. The effect of larval phase on susceptibility of the army worm, *Pseudaletia separata* Walker (Lepidoptera : Noctuidae) to nuclear polyhedrosis and granulosis viruses. *Proc. 18th Int. Nat. Conf. Ent., Canada*, pp. 261.

Kushwaha, K.S. and Bhardwaj, S.C., 1967. Biology and external morphology of forage pests. I. Tussock caterpillar *Euproctis* spp. (Lymantriidae : Lepidoptera). *Indian J. Agric. Sci.*, 39: 93–107.

Laing, D.R. and Caltagirone, L.E., 1969. Biology of *Habrobracon lineatellae. Can. Ent.*, 101: 135–142.

Lal, R. and Haque, E., 1955. Effect of nutrition under controlled conditions of temperature and humidity on longevity and fecundity of *Sphaerophora scutellaris* (Fabr.) (Syrphidae : Diptera). *Indian J. Ent.*, 17: 317–25.

Lal, L. and Mukharji, S.P., 1978. Growth potential of *Diacrisia obliqua* Walker in relation to certain food plants. *Indian J. Ent.*, 40: 177–181.

Lall, B.S., 1958. On the biology of *Apanteles obliquae* (Wlk) a larval parasite of *Diacrisia obliqua* (Wlk). *Indian J. Ent.*, 20: 291–295.

Lall, B.S., 1959. On the biology of *Apanteles obliquae* (Wlk) a larval parasite of *Diacrisia obliqua* (Wlk). *Curr. Sci.*, 20: 161–162.

Lall, B.S., 1964. Vegetable pests. In: *Entomology in India.* Ent. Soc. India, New India, p. 198.

Lefroy, H.M., 1907a. The more important insects injurious to Indian Agriculture. *Mem. Dept. Agric.* India, 1: 113–252.

Lefroy, H.M., 1907. The more important insect injurious to Indian Agriculture. *Mem. Dept. Agric.*, India, 1: 160.

Leong, G.K.L. and Oatman, E.R., 1968. The biology of *Campoplex haywardi* (Hymenoptera : Ichneumonidae), a primary parasite of the potato tuber worm. *Ann. Ent. Soc., Am.*, 61: 26–36.

LeRoux, E.J. and Others, 1963. Population dynamics of agricultural and forest insect pests. *Mem. Ent. Soc., Can.*, 32: 1–103.

Lewis, W.J. and Redlinger, L.W., 1969. Suitability of eggs of the almond moth, *Cadra cautella* of various ages of parasitism by *Trichogramma evanscens*. *Ann. Ent. Soc. Am.*, 62: 1482–1485.

Loan, C.C., 1963. Parasitism of the dogwood flea beetle, *Altica corni* in ontaria. *J. Econ. Ent.*, 56: 537–538.

Lotka, A.J., 1925. *Elements of Physical Biology*, Williams and Wilkins, Baltimore, Md, pp. 462.

Luckmann, W.H., 1963. Measurements of the incubation period of *Heliothis*. *J. Econ. Ent.*, 56: 60–62.

Madan, Y.P., Mrig, K.K. and Choudhary, J.P., 1987. Effects of some semi-artificial diets on the fecundity and longevity of *Pyrilla perpusilla* Walker adults. *Indian J. Ent.*, 49: 496–501.

Mally, P.W., 1892. Report of progress in the investigation of the cotton bollworm. *U.S. Dept. Agri. Div. Entomol. Bull.*, 26: 45–56.

Mani, M., 1985. Age specific fecundity and rate of increase of *Eucdatoria bryani* Sabrosky on *Heliothis armigera* (Hubn.). *Indian J. Ent.*, 47: 163–168.

Manoharan, T., Chockalingam, S. and Shanthi, P., 1984. Effect of food quality on development of *Euproctis fraterna* (Lymantriidae : Lepidoptera). *Comp. Physiol. Ecol.*, 9: 164–168.

Mason, W.R.M., 1981. The polyphyletic nature of *Apanteles* Foerster (Hymenoptera : Braconidae) a phylogeny and reclassificaton of microgastrinae, pp. 1–147.

Massodi, M.A., 1985. Growth response of *Lymantria obfuscata* Walker in relation to tannin content in different host foliages. *Indian J. Ent.*, 47: 422–426.

Masoodi, M.A. and Srivastava, A.S., 1985. Effect of host plants on the pupal weight and fecundity of *Lymantria obfuscata* Walker (Lymantriidae : Lepidoptera). *Indian J. Ent.*, 47: 410–412.

Mathew, G., Sudheendrakumar, V.V., Mohandas, H. and Nair, K.S.S., 1990. An artificial diet for the teak defoliator, *Hyblaea puera* Cramer (Lepidoptera : Hybiaeidae). *Entomon.*, 15: 159–164.

Mathur, A.C., 1962. Food-plant spectrum of *Diacrisia obliqua* Walk. (Arctiidae: Lepidoptera). *Indian J. Ent.*, 28: 278–279.

Mathur, Y.K., 1966. Cannibalism in *Amsacta moorei* Butler (Lepidoptera : Arctiidae). *Indian J. Entom.*, 30: 322–325.

Mathur, Y.K. and Mathur, B.L., 1968. Note on a parasitic *Aspergillus* on larvae of *Amsacta moorei* Butler. *Pl. Prot. Bull. FAD*, 16: 66.

Mathur, Y.M., Verma, J.P. and Sharma, S.K., 1971. *Microlestas discoidalis* Fairm. (Col: Carabidae) as a predator of *Eucoa segetum* Walk. (Lep : Noctuidae). *Entomologist's Monthly Magazine*, 10: 12.

Mech, L.D., 1974. A new profile for the wolf. *Natural History*, 83: 26–31.

McLeod, J.W., 1972. The Swaine Jack pine sawfly, *Neodiprion swalonei* life system. Evaluating the long term effect of insecticide application in Quebec. *Environ. Entomol. J.*, pp. 371–378.

Mellini, E., Galassi, L. and Brilini, G., 1979. Effectidella temperature sulla copia ospiteparassita *Galleria mellonella* L. *Gonia cinerascens* Rond. *Bollettino dell*

Instituto di Entomologia della Universita di Bologna 35: 13–28.

Messenger, P.S., 1964. Use of life tables in bioclimatic study of an experimental aphid braconid wasp host parasite system. *Ecology*, 45: 119–131 (W.L. 17406).

Mistry, A., Yadav, D.N., Patel, R.C. and Parmar, B.S., 1985. Field evaluation of nuclear polyhedrosis virus against *Heliothis armigera* Hubner (Lepidoptera : Noctuidae) in Gujarat. *Ind. J. Plant Proc.*, 12: 31–33.

Mohamad, A.K.A., Bell, J.V. and Sikorowshi, P.P., 1978. Field cage tests with *Nomuraea rilleyi* against corn earworm larvae on sweet corn. *J. Eco. Ent.*, 71: 102–104.

Morris, R.F. and Miller, C.S., 1954. The development of life tables for the spruce bud worm. *Can. J. Zool.*, 32: 283–301 (W.L.13191).

Nagarajan, K.R., Perumal, K. and Shanmugam, N., 1957. The red hairy caterpillar (*Amsacta albistriga* W.) and its field-scale control. *Madras Agric. J.*, 44: 150–153.

Nair, K.P.V. and Jacob, A., 1976. Studies on the nuclear polyhedrosis of *Pericallia ricini* F. (Lepidoptera : Arctiidae). *Entomon.*, 1: 23–30.

Nair, K.R., 1988. Field parasitism by *Apanteles flavipes* Cameron (Hymenoptera : Braconidae) on *Chilo partellus* (Swinh.) in *Coix lachryfmajobi* L. and *Chilo auricilius* (Dudgn.) in sugarcane in India. *Entomon.*, 13; 283–287.

Narayanan, E.S., 1959. Insect pests of castor. In: *Castor*, Indian Central Oilseeds Committee.

Narendran, T.C. and Joseph, K.J., 1976. Biological studies of *Brachymeria lasus* (Walker) (Hymenoptera : Chacidae). *Entomon.*, 1: 31–38.

Nataraju, B., Baig, Murthuza, Raju, Rajagopalan, Krishnaswami, S. and Samson, M.V., 1989. Feeding trials with different varieties of mulberry in relation to cocoon crop preference and incidence of loss due to diseases. *Indian J. Ent.*, 51(3): 238–241.

Nene, Y.L., 1973. Notes on a fungus parasite of *Bamisia tabaci* Gann, a vector of several plant viruses. *Indian J. Agri. Sci.*, 13: 514–516.

Nikam, P.K. and Sathe, T.V., 1981. Studies on the effect of temperature on the development of *Diadegma trichoptilus* (Cameron) (Hymenoptera : Ichneumonidae), an internal larval parasitoid of *Exelastis atomosa* Wals. *Marathwada Univ. J. Sci.*, 20: 37–38.

Nikam, P.K. and Sathe, T.V., 1983a. Life tables and intrinsic rates of natural increase of *Cotesia flavipes* (Cameron) (Hymenoptera : Braconidae) Population on *Chilo partellus* (Swin.) (Lep. Pyralidae). *Z. Ang. Ent.*, 95: 171–175.

Nikam, P.K. and Sathe, T.V., 1983b. Studies on host age selection by *Cotesia flavipes* (Cameron) a larval parasitoid of *Chilo partellus* (Swin.). *Indian J. Parasitol.*, 7: 181–182.

Nishida, T., 1956. An experimental study of the ovipositional behaviour of *Opius fletcheri* Silvestri, a parasite of a melonfly. *Proc. Hawaii Entomo.Soc.*, 16: 126–134.

Nixon, G.E.J., 1965. A reclassification of the tribe Microgasterini (Hymenoptera : Braconidae). *Bull. Br. Mus. Nat. Hist. (Ent.)*, 2: 1–284.

Nixon, G.E.J., 1967. The Indo-Australian species of the Ultor group of *Apanteles* Foerster (Hymenoptera : Braconidae). *Bull. Br. Mus. Nat. Hist. (Ent.)*, 21: 1–34.

Oatman, E.R., Plainer, G.R. and Greany, P.D., 1969. The biology of *Orgilus lepidus* (Hymenoptera : Braconidae), a primary parasite of the potato tuberworm. *Ann. Ent. Soc. Am.*, 62: 1407–1414.

Oatman, E.R., Hall, I.M.. Aradawa, K.Y., Platner, G.R., Bascom, L.A. and Beegle, C.C.., 1970. Control of the corn earworm on sweet corn in Southern California with a *Nuclear polyhedrosis* virus and *Bacillus thuringiensis. J. Econ. Entomol.*, 63: 415–421.

Oatman, E.R. and Platner, G.R., 1974. The Biology of *Temelucha* sp. *platensis* group (Hymenoptera : Ichneumonidae), a primary parasite of the potato Tuber worm. *Ann. Ent. Soc. Am.*, 67: 275–280.

Odebiyi, J.A. and Oatman, E.R., 1972. Biology of *Agathis gibbosa* (Hymenoptera : Braconidae), a primary parasite of potato tuber moth. *Ann. Ent. Soc. Am.*, 65: 1104–1114.

Pandey, S.N., 1968. Observations on biology of the castor hairy caterpillar, *Euproctis lunata* (Lepidoptera : Lymantriidae). *Indian J. Ent.*, 30: 263–265.

Pandey, S.N. and Srivastava, R.N., 1967. Growth of larvae of *Prodenia litura* F. in relation to wild food plants. *Indian J. Ent.*, 29: 229–233.

Pandey, N.D., Yadav, D.R. and Teotia, T.P.S., 1968. Effect of different food plants on larval and post larval development of *D. obliqua* (Walker). *Indian J. Ent.*, 30: 221–234.

Paripurna, K. and Srivastava, B.G., 1989. Effect of different quantities of sucrose and glucose on the growth and development of *Dacus cucurbitae* (Coquillett) maggots under aseptic condition. *Indian J. Ent.*, 51: 229–233.

Parker, H.L., 1924. Recherches Sur les formes Postembryonaires des Chalcidiens. *Ann. Soc. Ent.*, *France*, 93: 261–393.

Patel, H.K. and Patel, V.C., 1965. Life history, epidemiology and seasonal history of Gujarat hairy caterpillar (*Amsacta moorei* Butler). *Bansilal and Amritlal Coll. Mag.*, 16 and 17: 44–56.

Patel, A.G. and Vyas, H.N., 1984. Studies of predatory capacity of lady bird beetle, *Manochilus saxmaculatus* Fabricius against *Aphis craccivora* Hoch. under laboratory conditions. *Pesticides*, 18: 8–9.

Patel, J.R. and Jotwani, M.G., 1986. Effect of ecological factors on incidence and damage by sorghum midge, *Contarinia sorghicola*. *Indian J. Ent.*, 48: 220–222.

Patel, R.M., Patel, C.B. and Vyas, H.N., 1978. Record and some observations on local milky disease in white grubs *Helotrichia* spp, near *consaquinea* Blanch in India. *Indian J. Ent.*, 39: 181–182.

Patil, V.S and T.V. Sathe., 2003. *Insect Predators and Pest Management*. Daya Publishing House, Delhi, pp. 1–216.

Pawar, C.S. and Jadhav, D.R., 1983. Wasps-predators of *Heliothis* on pigeonpea. *International pigeonpea Newsletter ICRISAT*, 2: 65–66.

Philipp, J.S. and Watson, T.F., 1971. Influence of temperature on population growth of the pink bollworm. *Ann. Ent. Soc. Am.*, 64: 334–340.

Polis, G.A., 1981. The evolution and dynamics of intraspecific predation. *Ann. Rev. Ecol. Syst.*, 12: 225–251.

Poonia, F.S., 1985. Consumption, digestion and utilization of castor leaves by larvae of Eri silkworm, *Philosamia ricini* Hutt (Saturniidae : Lepidoptera). *Indian J. Ent.*, 47: 255–267.

Pruthi, H.S., 1969. *Text book of Agricultural Entomology*. Indian Council of Agricultural Research, New Delhi, pp. 977.

Quednau, F.W. and Guevermont, H., 1975. Observation on mating and oviposition behaviour of *Priopoda nigricollis* (Hymenoptera : Ichneumonidae), a parasite of brich leaf miner, *Fenusa pusilla* (Hymenoptera : Tenthredinidae). *Can. Ent.*, 107: 1199–1204.

Rabindra, R.J. and Subramaniam, T.R., 1974. A nuclear polyhedrosis of *Dasychira mendosa* Hb (Lepi : Lymantriidae). *Curr. Sci.*, 43: 721–722.

Rabindra, R.J., Paul, A.V.N., David, B.V. and Subramaniam, T.R., 1975. On the nuclear polyhedrosis of *Plusia chalcytes* Esp. (Lep : Noctuidae). *Curr. Sci.*, 44: 273–274.

Raj, B.T., 1988. Seasonal variation in the male population of potato tubermoth, *Phthorimaea operculella* (Zeller) in the Deccan plateau. *Indian, J. Ent.*, 50: 24–27.

Rajasekhara, K., Chatterji, S. and Ramdas Menon, M.G., 1964. Biological notes on *Psallus* sp. (Miridae : Hemiptera), a predator of *Taeniothrips nigricornis* Schmutz (Thripidae : Thysanoptera). *Indian J. Ent.*, 26: 62–66.

Rajasekhara, K. and Chatterji, S., 1970. Biology of *Orius indicus* (Hemiptera : Anthocoridae), a predator of *Taeniothrips nigricornis* (Thysanoptera). *Ann. Ent. Soc., Am.*, 63: 364–367.

Rajendra, M.K. and Patel, R.C., 1971. Studies on the life history of a predatory pentatomid bug, *Andrallus spinidens* (Fabr.). *J. Bombay Nat. His. Soc.*, 68: 319–327.

Ramaseshiah, G., 1973. *Entomophthora grylli* Fres. on Arctiid larvae in India. *Tech. Bull. Corn. Inst. Biol. Cont.*, 16: 35–39.

Ramakrishnan, N., Pawar, V.M. and Prakash, N., 1975. Nuclear polyhedrosis of *Spodoptera litura* (Fabr.) description of inclusion bodies and virions. *Curr. Sci.*, 44: 316–317.

Rao, S.N., 1961. Key to the oriental species of *Apanteles Foerster* (Hymenoptera). *Proc. Nat. Acad. Sci., India,* (B)31: 32–46.

Rao, V.P., 1969. Survey for natural enemies of aphids in India. *Final tech. rep; Inst. Biol. Control. Indian Stn.,* Bangalore, India, pp. 1–93.

Rao, P.S., 1980. A new record of predatory wasp, *Polistes hebreaus* Fabricious associated with the pest complex of green gram, *Vigna radiata* Wilczek. *Indian J. Ent.*, 42: 278–279.

Rathore, Y.S. and Sachan, G.C., 1976. Developmental behaviour of *Diacrisia obliqua* (Walker) (Lepidoptera : Arctiidae) on some common weeds. *Curr. Sci.*, 47: 104–106.

Rathore, Y.S. and Sachan, G.C., 1981. Use of D. statistics and canonical variate analysis in assessing the developmental potential of *Diacrisia obliqua* (Walker) on ornamental plants. *Indian J.Hort.*, 38: 268–274.

Reddy Veera, C.G. and Bhattacharya, A.K., 1988. Life cycle of *Heliothis armigera* (Hubner.) on semisynthetic diets. *Indian J. Ent.*, 50: 357–370.

Richards, O.W. and Waloff, N., 1961. A study of a natural population of *Phytoclecta olivacea* (Foerster) (Coleoptera : Chrysomelidae). *Trans. Roy. Entomol. Soc., London*, B. 244: 205–257.

Rojas-Rousse, D. and Benoit, M., 1977. Morphology and biometry of larval instars of *Pimpla instigator* (F) (Hymenoptera : Ichneumonidae). *Bull. Ent. Res.*, 67: 129–141 (W.L. 10184).

Root, R.B., 1973. Organization of a plant-arthropod association in simple and diverse habitats, the fauna of colards (*Brassica oleracea*). *Ecol. Monogr.*, 42: 95–124.

Santhakumar, M.V., 1989. Behavioural studies of some hymenopterous parasitoids on lepidopterous pests of chickpea and pigeonpea in relation to reproduction. *Ph.D. Thesis*, Shivaji University, Kolhapur, pp. 1–223.

Sathe, T.V., 1985a. Development and survival of *Cotesia diurnii* R. and N., a larval parasitoid of *Exelastis atomosa* Wals. in relation to different constant temperatures. *Geobios*, 12: 46–47.

Sathe, T.V., 1985b. Studies on mating, oviposition and emergence of *Cotesia diurnii* R. and N. (Hymenoptera : Braconidae), an internal larval parasitoid of *Exelastis atomosa* Wals. *Geobios New Reports,* 4: 100–101.

Sathe, T.V., 1986a. Life table and intrinsic rate of increase of *Cotesia diurnii* Rao and Nikam (Hymenoptera : Braconidae), a larval parasitoid of *Exelastis atomosa* Wals. *Entomon.,* 11: 281–283.

Sathe, T,V., 1986b. Biology of *Cotesia diurnii* R. and N. (Hym. : Braconidae), a larval parasitoid of *Exelastis atomosa* Walsingham. *Oikoassay,* 3: 31–33.

Sathe, T.V., 1987. Longevity, fecundity and sex-ratio of *Diadegma trichoptilus* (Cam.) (Hym : Ichneumonidae), a larval parasitoid of *Exelastis atomosa* Wals., *Geobios,* 14: 268–269.

Sathe, T.V., 1988a. Intrinsic rate of increase and interspecific relationship between *C. orientalis, C. diurnii* and *D. tricoptilus,* larval parasitoids of *E. atomosa* Walsingham. *Advances in Parasitic Hymenoptera Research,* USA, pp. 383–387.

Sathe, T.V., 1988b. The biology of *Cotesia orientalis* C. and N. (Hymenoptera : Braconidae). *J. Zool. Res.,* 1: 23–27.

Sathe, T.V., 1990. The biology of *Diadegma argenteopilosus* Cameron (Hymenoptera : Ichneumonidae), an internal larval parasitoid of *Spodoptera litura* (Fab.). *The Entomologist,* 109: 2–7.

Sathe., T.V., 2004. *Vermiculture and Organic Farming.* Daya Publishing House, Delhi, pp. 122.

Sathe, T.V. and Bosale, Y.A., 2001. Insect Pest Predators. Daya Publishing House, Delhi, pp. 1–169.

Sathe, T.V. and Nikam, P.K., 1983a. Adult longevity of *Cotesia flavipes* (Cameron) (Hymenoptera : Braconidae), with different food. *Sci. and Cult.* 49: 405–406.

Sathe, T.V. and Nikam, P.K., 1983b. Mating, oviposition and emergence of *Diadegma trichoptilus* (Cameron) (Hymenoptera : Ichneumonidae), a larval parasitoid of *Exelastis atomosa* Wals. *Curr. Sci.,* 52: 501–502.

Sathe, T.V. and Nikam, P.K., 1984. Mating, oviposition and emergence of *Cotesia orientalis* C. and N. (Hym : Braconidae), an internal larval parasitoid of *Exelastis atomosa* Wals. *Comp. Physiol. Ecol.,* 9: 231–232.

Sathe, T.V. and Nikam, P.K., 1985. Influence of certain dietary combinations on longevity of adults of *Diadegma trichoptilus* (Cameron), a larval parasitoid of *Exelastis atomosa* Walsingham. *Geobios,* 12: 1964–66.

Sathe, T.V. and Nikam, P.K., 1986. Biology and Biometry of *Diadegma trichoptilus* (Cameron) (Hym : Ichneumonidae), a larval parasitoid of *Exelastis atomosa* Walsingham. *Marath. Univ. J. Sci.,* 17–18: 61–66.

Sathe, T.V., Inamdar, S.A. and Ingawale, D.M., 1988a. Cannibalism in *Heliothis armigera* (Hubn.) (Lepidoptera : Noctuidae). *Oikoassay,* 5: 1.

Sathe, T.V., Inamdar, S.A. and Dawale, R.K., 2003. *Indian Pest Parasitoids*. Daya Publishing House, Delhi, pp. 197.

Sathe, T.V., Santhakumar, M.V. and Inamdar, S.A., 1988b. Biology of *Apanteles creatonoti* Viereck (Hymenoptera), a larval parasitoid of *Thiocidas postica* Wlk. (Lepidoptera). *Entomon.,* 13: 189–190.

Sato, Y., 1975. Rearing *Apanteles glomeratus* L. on the larvae of *Pieris rapae crucivora* Boisdual fed on an artificial diet. *Kontyu* (Tokyo), 43: 242–249 (W.L. 27647).

Searle, T. and Yule, W.N., 1988. Fungal control of the carrot weevil, *Listronotus oregonensis*. *Proc. 17th Int. Nat. Cong. Ent.*, Canada, pp. 262.

Sen, P. and Mukherjee, A.B., 1955. Preliminary note on the life history of *Amsacta lactinea* (Cram), a pest of groundnut (*Arachis hypogaea*). *Proc. 42nd Indian Sci. Congr.*, 3: 291.

Sethi, G.R., Singh, K.M. and Prasad, H.H., 1976. Upsurge of Bihar hairy caterpillar, *Diacrisia obliqua* Walker on sunflower. *Entomologists Newsletter*, 6: 35.

Sharma, J.C., 1975. Development of *Menochilus sexmaculata* Fab. as influenced by feeding on different species of aphid hosts. *JNKVV Res. J.*, 8: 275.

Sharma, S.K. and Chaudhary, J.P., 1985. Effect of adult nutrition on the reproductive behaviour of *Heliothis armigera* (Hubner). *Indian J. Ent.*, 47: 433–436.

Sharma, S.K. and Chaudhary, J.P., 1988. Effect of different levels of constant temperature and humidity on the development and survival of *Heliothis armigera* (Hubner). *Indian J. Ent.*, 50: 76–81.

Short, J.R.T., 1959. A description and classification of final instar larvae of Ichneumonidae (Insecta : Hymenoptera). *Proc. U.S. Nat. Mus.*, 110: 391–511.

Short, J.R.T., 1970. On classification of the final instar larvae of the Ichneumonidae (Hymenoptera). *Suppl. Trans. R. Ent. Soc. Lond.*, 122: 185–210 (W.L. 53980).

Singh, Z., 1975. New record of a predatory bug *Amyecea* (*Asopus*) *malabarica* (Fabr.) on *Nezara viridula* Linn. in India. *Indian J. Ent.*, 35: 169–170.

Singh, J. and Butter, N.S., 1985. Influence of climatic factors on the build up of whitefly, *Bemisia tabaci* Genn. on cotton. *Indian J. Ent.*, 47: 359–360.

Singh, O.P. and Gangrade, G.A., 1974. Biology of *Diacrisia obliqua* Walker (Lepidoptera : Arctiidae) on soyabean and effect of loss of chlorophyll on pod grain. *JNKVV Res. J.*, 8: 86–91.

Singh, O.P. and Gangrade, G.A., 1975. Parasites, predators and diseases of larvae of *Diacrisia obliqua* Walker (Lepid : Arctiidae) on soyabean. *Curr. Sci.*, 44: 481–482.

Singh, H. and Ram, B., 1987. Effect of different host plants on the development of red cotton bug, *Dysdercus koenigii* (Fab). *Indian J. Ent.*, 49: 345–350.

Singh, H., Singh. Z. and Naresh, J.S., 1986. Path coefficient analysis of abiotic factors affecting the aphid populations on rape seed. *Indian J. Ent.*, 48: 156–161.

Singh, K.N. and Sachan, G.C., 1987. Development behaviour of *Diacrisia obliqua* (Walker) (Lepidoptera : Arctiidae) on sugarbeet. *Indian J. Ent.*, 49: 429–435.

Singh and Mavi, G.S., 1986. Growth and development of leaf miner, *Phytomyza horticola* Goureau on some *Brassica* hosts. *Indian J. Ent.*, 48: 301–304.

Singh, S.V. and Singh, Y.P., 1989. Effect of insecticides on aphid population, plant growth and yield of mustard crop. *Indian J. Ent.*, 51: 11–18.

Singh, T., Vinod Kumar and Reddy, G.P.V., 1987. Nutritional requirements of the castor semilooper. *Achaea janata* Linn. *Indian J. Ent.*, 49: 502–514.

Sinha, R.P., Yazdani, S.S. and Verma, G.D., 1989. Population dynamics of mustard aphid, *Lipaphis erysimi* Kalt. in relation to ecological parameters. *Indian J. Ent.,* 51: 334–339.

Sinha, P.P.. Yazdani, S.S. and Verma, G.D., 1990. Population dynamics of mustard aphid *Lipaphis erysimi* Kalt. (Homoptera : Aphididae) in relation to ecological parameters. *Indian J. Ent..* 52: 387–392.

Smilowitz, Z. and Iwantsch, G., 1975. Relationship between the parasitoid, *Hyposoter exigua* and cabbage looper *Trichoplusia* ni. The effect of host age on ovipositional rate of the parasitoid and successful parasitism. *Can. Ent.,* 107: 689–694.

Srivastava, A.S., 1964. Entomological research during the last ten years in the section of the Entomologist of Government of U.P. Kanpur (India). *Res. Mem., Uttar Pradesh,* 3: 1–56

Srivastava, A.S. and Goel, B.K., 1962. Bionomics and control of red hairy caterpillar. *Amsacta moorei. Bull. Proc. Nat. Acad. Sci., India,* B32: 97–100.

Srivastava, A.S., Katiyar, R.R., Upadhaya, K.D. and Singh. S.V., 1987. Studies on the food preference of *Coccinella repanda* Thunberg (Coleoptera : Coccinelidae). *Indian J. Ent.,* 52: 551–552.

Srivastava, K.M. and Pant, J.C., 1989. Growth and developmental response of *Callosobruchus maculata* (Fabr.) to different pulses. *Indian J. Ent.,* 51: 269–272.

Srivastava, A.S., Siddiqi, M.S. and Saxena, H.P., 1965. Bionomics and control of *Amsacta moorei* Bull. (Lepidoptera : Arctiidae) with reference to allied species. *A. lactinea* Cram. *Labdev J. Sci. Technol.,* 3: 26–29.

Stark, R.W., 1959. Population dynamics of the lodgepole needle miner. *Necrophorus* spp. and the mite. *Poecilochirus necrophori* Vitz. *J. Anim. Ecol.*, 37: 417–424 (W.L. 25559).

Stinner, R.E., Rubb, R.I. and Bradley, J.R., 1976. Natural factors operating in the population dynamics of *Heliothis zea* in North Carolina. *Proc. XV Int. Compr. Entomol.*, 622–640.

Sweetman, H.L., 1958. *The Principles of Biological Control.* Dubuque, Wm. C. Brown 8.

Sychevskaya, V.I., 1966. Biology of *Brachymeria minuta* (Hymenoptera : Chalcidoidea), a parasitoid of synanthropic flies of the family sarcophagidae (Diptera). *Ent. Rev.*, 45: 424–429.

Talati, G.M. and Butani, P.G., 1979. Predatory capacity of *Coccinella septampunctata* on groundnut aphid, *Ind. J. Prot.*, 7: 107.

Thobbi, V.V., 1961. Growth potential of *Prodenia litura* F. in relation to certain food plants. *Indian J. Ent.*, 23: 262–264.

Thobbi, V.V. and Singh, B.U., 1974. First record of *H. armigera* as a predator on the pupae of castor semilooper, *Achaea janata* (L.). *Curr. Sci.*, 43: 354–355.

Thompson, W.R., 1924. La theoric mathematique deiaction des parasites entomophages et le facteur du hazard. *Annls. Fac. Sci. Marseilla*, 2: 69–89.

Thorpe, W.H., 1932. Experiments upon respiration in the larvae of certain parasitic Hymenoptera. *Proc. Roy. Entomol. Soc., London*, (B) 109: 450–471.

Thurston, R. and Postley, L., 1978. Effect of instar of *Manduca sexta* on the rate of development of its parasite, *Apanteles congregatus*. *Tob. Sci.,* 22: 32–34.

Tikar, D.T. and Thakare, K.P., 1961. Bionomics, biology and immature stages of an Ichneumonid, *Horogenus fenestralis* Holmgren, a parasite of common caterpillar. *Indian J. Ent.,* 23: 116–124.

Tiwari, S.N., Rathore, Y.S., Bhattacharya, A.K. and Sachan, G.C., 1988. Susceptibility of several varieties of groundnut to *Spilosoma obliqua* Walker (Lepidoptera : Arctiidae). *Indian J. Ent.,* 50: 179–184.

Townes, H., Townes, M. and Gupta, V.K., 1961. A catalogue and reclassificatlon of the Indo-Australian Ichneumonidae, pp. 522. *Mem. Am. Ent. Inst.,* 1: 1–522.

Vance, A.M. and Smith, H.D., 1933. The larval head of parasitic Hymenoptera and nomenclature of its parts. *Ann. Ent. Soc., Am.,* 26: 86–94.

Vasantharaj, D.B. and Janagarajan, A., 1966. *Dactynotus carthami* H.R.L. a new host for the predatory coccinellid *Brumus suturalis* Fabr. *Madras Agric. J.,* 53: 295.

Verma, A.K. and Makhmoor, H.D., 1988. The intrinsic rate of natural increase of the cabbage aphid, *Brevicoryne brassicae* (Linn.) (Homoptera : Aphididae) on Cauliflower. *Entomon.,* 13: 51–55.

Vijayalakshmi, K., 1986. Predatory efficiency in giant crab spider-cockroach system: Predatory behaviour and biocontrol. *Proc. Orient. Ent. Sym.,* 11: 133–140.

Waloff, N., 1968. Studies on the insect fauna on scotch broom *Sarothammus scoparius* (L.) Wimmer. *Adv. Ecol. Res.,* 5: 87–208.

Watson, T.F., 1964. Influence of host plant conditions on population increase of *Tetranychus telarius* (Linnaeus) Hilgardia, 35: 273–322.

Watson, T.F. and Johnson, P.H., 1972. Life cycle of the cotton leaf-perforator. *Prog. Agn. Ariz.* 24: 12–13.

Weseloh, R.M., 1977. Mating behaviour of Gypsy moth parasite, *Apanteles melanoscelus. Ann. Ent. Soc. Am.,* 70: 549–554.

Whitecomb, W.H., 1973. Natural populations of entomophagous antropods and their effect on the agro-ecosystem. In: *Proc. Summer Inst. on Biol. Control of Plant Insects and Diseases.* Miss. State Univ. June 19–30, 1972 pp.

Yadava, C.P. and Lal, S.S., 1988. Relationship between certain abiotic and biotic factors and the occurrence of gram pod borer, *Heliothis armigera* (Hubn.) on Chickpea. *Entomon.,* 13: 269–273.

Yadava, P.P., Singh, R., Kumar, A.K., Mahto, D.N. and Sinha, M.M., 1978. Larval survival and growth potential of *Diacrisia obliqua* Walker in relation to some promoting varieties. *Indian J. Ent.,* 40: 146–149.

Yeargan, K.V. and Latheef, M.A., 1977. Ovipositional rate, fecundity and longevity of *Bathyplectes anurus,* parasite of the alfalfa weevil. *Environ. Entomol.,* 6: 31–34.

Yeargan, K.V., Parr. J.C. and Pass, B.C., 1978. Fecundity and longevity of *Bathyplectes curculionis* under constant and fluctuating temperatures. *Environ. Entomol.,* 7: 36–38.

Index

www.ingramcontent.com/pod-product-compliance
Lightning Source LLC
Chambersburg PA
CBHW072250210326
41458CB00073B/922